OUR COSMIC HABITAT

MARTIN REES

OUR COSMIC HABITAT

Weidenfeld & Nicolson
LONDON

First published in the USA in 2001 by
Princeton University Press, 41 William Street,
Princeton, New Jersey 08540

First published in Great Britain in 2002 by
Weidenfeld & Nicolson

A CIP catalogue reference for this book
is available from the British Library

ISBN 0 297 82901 7

Printed in Great Britain by
Butler & Tanner Ltd, Frome and London

Weidenfeld & Nicolson

The Orion Publishing Group Ltd
Orion House
5 Upper Saint Martin's Lane
London, WC2H 9EA

I PERCEIVED that I was on a little round grain of rock and metal, filmed with water and with air, whirling in sunlight and darkness. And on the skin of that little grain, all the swarms of men, generation by generation, had lived in labour and blindness, with intermittent joy and intermittent lucidity of spirit. And all their history, with its folk-wanderings, its empires, its philosophies, its proud sciences, its social revolutions, its increasing hunger for community, was but a flicker in one day of the lives of the stars.

OLAF STAPLEDON
Star Maker (1937)

CONTENTS

It was a privilege to be asked to speak on "Our Cosmic Habitat" in the first Scribner Lectures, a new annual series co-sponsored by Princeton University and Princeton University Press. But it was daunting as well, because Princeton is one of the world's leading centers for research in cosmology, and I normally go there to learn rather than to talk.

One of Princeton's most eminent and inspirational scientists, Professor John A. Wheeler, taught me the neat aphorism, "Time is Nature's way of stopping things happening all at once." I offer, in return, a more frivolous one: "God invented space so that not everything had to happen in Princeton."

Nonetheless, much of the action in cosmology did happen there, and it was perhaps foolhardy of me to choose a theme that had greater experts locally. However, my lectures weren't addressed to them: my aim was to offer a broad-brush picture of some lively scientific frontiers, emphasizing new ideas in a way that was accessible to a general audience.

In this written text, I have adjusted the balance of topics, highlighting points that are less familiar to most, and perhaps in consequence more speculative.

The provost of Princeton University, Jeremiah Ostriker, kindly invited me to give the Scribner Lectures. I thank him, and other friends and colleagues, especially J. Richard Gott, for their hospitality and support. I am also grateful to Walter Lippincott, Trevor Lipscombe, and Fred Appel at the University Press for their practical help during my visit, to Alice Calaprice and Joe Wisnovsky for editorial advice in preparing this text, and to Richard Sword for drawing the figures.

Could God Have Made the World Any Differently?

The preeminent mystery is why anything exists at all. What breathes life into the equations of physics, and actualized them in a real cosmos? Such questions lie beyond science, however: they are the province of philosophers and theologians. For science, the overarching problem is to understand how a genesis event so simple that it can be described by a short recipe seems to have led, 13 billion years later, to the complex cosmos of which we are a part. Was the outcome "natural," or should we be surprised at what happened? Could there be other universes? Scientists are now addressing such questions, which had formerly been in the realm of speculation. Cosmology has a history that stretches back for millennia, but the conceptual excitement has never been more intense than it is at the start of the twenty-first century.

The Sun and the firmament are part of our environment—our cosmic habitat. Artistic and mystical geniuses share this perception with scientists. D. H. Lawrence wrote, "I am part of the Sun as my eye is part of me." Van Gogh's "Starry Night" was painted in the same spirit as his pictures of cornfields and sunflowers. One can find numerous other such examples in the arts.

Science deepens our sense of intimacy with the nonterrestrial. We are ourselves poised between cosmos and microworld. It would take as many human bodies to make up the Sun's mass as there are atoms in each of us. Our existence depends on the propensity of atoms to stick together and to assemble into the complex molecules in all living tissues. But the atoms of oxygen and carbon in our bodies were themselves made in faraway stars that lived and died billions of years ago.

Technical advances during the twentieth century, especially its later decades, have enriched our perspective on our cosmic habitat. Space probes have beamed back pictures from all the planets of our solar system: new technology enables a worldwide public to share this vicarious cosmic exploration. Pictures of a comet crashing into Jupiter, made with the Hubble Space Telescope, were viewed almost in real time by more than a million people on the Internet. During this first decade of the twenty-first century, probes will trundle across the surface of Mars and even fly over it; they will land on Titan, Saturn's giant moon; and samples of Martian soil may be collected and brought back to Earth.

Our universe extends millions of times beyond the remotest stars we can see—out to galaxies so far away that their light has taken 10 billion years to reach us. Bizarre cosmic objects—quasars, black holes, and neutron stars—have entered the general vocabulary, if not the common understanding. We have learned that most of the stuff in the universe is not at all in the form of ordinary atoms: it consists of mysterious dark particles, or energy that is latent in space. We now envision our Earth in an evolutionary context stretching back before the birth of our solar system—right back, indeed, to the primordial event that set our entire cosmos expanding from some entity of microscopic size.

Deeper insight into the nature of space and time may enlarge our conception of the cosmos to embrace other universes beyond our own. These may manifest extra spatial dimensions and other concepts so far from our intuition that we shall grasp them with difficulty, if at all. What is surely astounding is that this enterprise has made any headway at all.

The public image of Albert Einstein is not the single-minded and ambitious researcher of his creative youth, but the benign and unkempt sage of his later Princeton years. One of the most-quoted of his aphorisms is: "The most incomprehensible thing about the universe is that it is comprehensible." He was here expressing his amazement that the laws of physics, which our minds are somehow attuned to understand, apply not just here on Earth, but everywhere we look. Our universe could have turned out to be an anarchic place, where atoms and the forces governing them are bafflingly different elsewhere in the cosmos from those we can study locally. But atoms in the most distant galaxies seem identical to those in our laboratories. Without this simplifying feature, we would have made far less progress in understanding our cosmic environment.

But what about the many things that remain incomprehensible? The most daunting challenge is posed by our biosphere—the immense complexity and variety of organisms, ecosystems, and brains. My interest lies in issues that I genuinely think are more tractable: probing and constraining the underlying laws that govern the microworld of atoms and the grand scale of the cosmos, and understanding how these set the stage for life by allowing the emergence of planets, stars, and galaxies.

In the last few years of the twentieth century, an exciting new research area opened up: the detection of planets around other stars. The night sky will soon be far more interesting. Stars will not be just points of light: many will have a distinctive retinue of planets whose main properties we will know. Will any of these harbor intelligence—or, indeed, even the most primitive life?

If aliens exist, and if we ever establish contact with them, what common culture might we share? The obvious answer is: our cosmic habitat. However different their evolution, the aliens would be made of atoms and governed by the same forces that govern us. If they had eyes and their world had clear skies, they would gaze out on the same vista of stars and galaxies that surround us. We and they would be confronted with stupendous expanses of space, as well as huge spans of time. Contemplative aliens might already have answered questions such as: What happened before the Big Bang? What causes gravity and mass? Is the universe infinite? How did atoms assemble—on at least one planet around at least one star—into beings able to ponder these mysteries? These questions still baffle all of us. Rather than the "end of science" being nigh, we are still near the beginning of the cosmic quest.

To link cosmos and microworld requires a breakthrough. Twentieth-century physics rests on two great foundations: the quantum principle (governing the "inner space" of atoms) and Einstein's relativity theory, which describes time, outer space, and gravity but doesn't incorporate quantum effects. The structures erected on these foundations are still disjoint. Until there is a unified theory of the forces governing both cosmos and microworld, we won't be able to understand the fundamental features of our universe: these features were imprinted on it at the very beginning, when everything was so squeezed that quantum fluctuations could shake the entire universe.

In his later life, Einstein focused on deep issues that are likely to attract more interest in the twenty-first century than they ever did in the twentieth. He spent his last thirty years in a vain (and, with hindsight, premature) quest for a unified theory of physics. Will such a theory—reconciling gravity with the quantum principle and transforming our conception of space and time—be achieved in coming decades?

The smart money is on a concept known as "superstring theory," or M-theory, in which each point in our ordinary space is actually a tightly folded origami in six extra dimensions, wrapped up on scales perhaps a billion billion times smaller than an atomic nucleus, and particles are represented as vibrating loops of "string." There is still an unbridged gap

MARTIN REES

between this elaborate mathematical theory and anything we can actually measure. Nonetheless, its proponents are convinced that string theory has a resounding ring of truth about it and that we should take it seriously.

A universe hospitable to life—what we might call a *biophilic* universe—has to be very special in many ways. The prerequisites for any life—long-lived stable stars, a periodic table of atoms with complex chemistry, and so on—are sensitive to physical laws and could not have emerged from a Big Bang with a recipe that was even slightly different. Many recipes would lead to stillborn universes with no atoms, no chemistry, and no planets; or to universes too short lived or too empty to allow anything to evolve beyond sterile uniformity. This distinctive and special-seeming recipe seems to me a fundamental mystery that should not be brushed aside merely as a brute fact.

How we respond to this mystery will depend on the answer to another of Einstein's questions: "Could God have made the world any differently?" Our universe, along with the physical laws that prevail in it, may turn out to be the unique outcome of a fundamental theory—in other words, nature may allow only one recipe for a universe. Alternatively, the underlying laws could be more permissive: they may allow many recipes, leading to many different universes; and these universes may actually exist.

We do not know which one of these options prevails. The answer will have to await a successful fundamental theory, and it would be presumptuous to prejudge the answer. Nonetheless, this book will focus on the fascinating consequences of the answer to Einstein's question—posed as the title of this Prologue—being "yes": God did have a choice. The entity traditionally called the universe—the entire domain that astronomers study, or the aftermath of "our" Big Bang—would be just one small element, or atom, in an infinite and immensely varied ensemble. The entire "multiverse" would be governed by a set of fundamental principles, but what we call the laws of nature would be no more than local bylaws—the outcome of historical accidents during the initial instants after our own particular Big Bang.

In this book I argue that the multiverse concept is already part of empirical science: we may already have intimations of other universes, and we could even draw inferences about them and about the recipes that led to them. In an infinite ensemble, the existence of some universes that are seemingly fine-tuned to harbor life would occasion no surprise; our own cosmic habitat would plainly belong to this unusual subset. Our entire universe is a fertile oasis within the multiverse.

PART I From Big Bang to Biospheres

1 Planets and Stars

The Sun

"Whilst this planet has been cycling on according to the fixed law of gravity, from so simple a beginning, forms most wonderful . . . have been and are being evolved." These are the famous closing words of Charles Darwin's *On the Origin of Species.*

Darwin's genius was to recognize how "natural selection of favored variations" could have transformed primordial life (formed, he surmised, in a "warm little pond") into the amazing varieties of creatures that crawl, swim, or fly on Earth. But this emergence—a higgledy-piggledy process, proceeding without any guiding hand—is inherently very slow. Darwin guessed that it would require hundreds of millions of years. This expanse of time didn't seem to faze him, because geologists had already invoked such timespans to account for the laying-down of rocks and the shaping of the Earth's surface features. Indeed, in the first edition of his book, Darwin estimated the rate of erosion in the Weald of Kent, a broad valley near his home, and gauged that this geological feature had to be 300 million years old. He insisted that "there is a grandeur in this view of life." To his nineteenth century contemporaries, these huge timescales were themselves mind-stretching compared to the constricted timespans of more traditional Western cosmologies.

There were, however, some troubling arguments that seemed to preclude such an ancient Earth. Lord Kelvin, one of the most celebrated physicists of his time, calculated that it would take only a few million years for the heat stored in Earth's molten core to leak out. And the Sun itself, he claimed, was radiating away its internal heat so fast that it would deflate in 10 million years. Kelvin's views carried great weight: "We may say, with certainty," he wrote, "that the inhabitants of the Earth cannot continue to enjoy the light and heat essential to their life, for many millions of years longer, unless sources now unknown to us are prepared in the great storehouse of creation."[1] The American geologist Thomas Chamberlain retorted that "there is perhaps no beguilement more insidious and dangerous than an elaborate and elegant mathematical process built upon unfortified premises."

Chamberlain was writing in 1899, but his words strike a

special resonance today. In later chapters, I shall discuss some intellectually alluring theories that may account for the most basic features of our physical universe—why it is expanding the way it is, why it contains atoms (and other particles), the strengths of the forces between them, and the nature of space itself. But the boldest and most ambitious of these are still based on "unfortified premises."

Even before Kelvin's death, advances in physics eroded the credibility of his estimates. Henri Becquerel's discovery in 1896 that uranium emitted mysterious rays gave the first hint of hitherto unenvisioned energy latent within atoms, and of what the Sun's actual fuel might be. Moreover, it was realized that power generated by radioactive decays in Earth's core could replenish the heat that was leaking out. But identifying the precise reactions that allowed the Sun to shine for 10 billion years as a gravitationally confined fusion reactor had to await a better understanding of atomic nuclei.

By the 1930s, it was realized that an atom consisted of a positively charged nucleus, surrounded by orbiting electrons with negative electric charge. The atoms of the different elements were distinguished from one another by the charge on their nucleus and the number of electrons needed to neutralize it. This number is 1 for hydrogen and 92 for uranium, the heaviest naturally occurring atom. The chemical properties of the various elements were already known in the nineteenth century, when the great Russian chemist Dmitri Mendeleyev put forward the periodic table of the elements, which showed certain patterns and family resemblances among different groups of elements. During the early twentieth century, the new quantum theory offered a natural explanation for these patterns, showing how they depended on the detailed orbits of the electrons surrounding the nuclei of atoms.

The atomic nuclei themselves are made up of protons and neutrons: the number of protons determines the electric charge and the place in the periodic table. A nucleus of helium, the second simplest element, consists of two protons and two neutrons. It weighs 0.7 percent less than the four particles from which it is made, however, and this difference in mass corresponds (via Einstein's famous equation $E = mc^2$) to the energy released when hydrogen is transmuted into helium. Nuclear fusion releases about a million times more energy per kilogram than any chemical process—combustion or explosion. The violence of an H-bomb testifies to fusion's power. Chemical processes merely alter or reshuffle the orbiting electrons; on the other hand, nuclear fusion taps a far larger reservoir of energy in the nuclei themselves.

The first calculation of how this fusion occurs in a star like the Sun depended on the insights of three great physicists. The first was a Russian, George Gamow, who will reappear in chapter 5 as one of the pioneers of the Big Bang cosmological theory. Gamow used the newly formulated quantum theory to estimate how hot the Sun's core would need to be for hydrogen fusion to be self-sustaining. The detailed chain reactions were worked out by Carl-Friedrich von Weizsaecker and Hans Bethe. These two eminent (and then young) physicists were both Germans, though Bethe went to the United States in the 1930s, having already achieved distinction as a pioneer of nuclear physics; amazingly, he still remained active in the subject at the beginning of the twenty-first century.

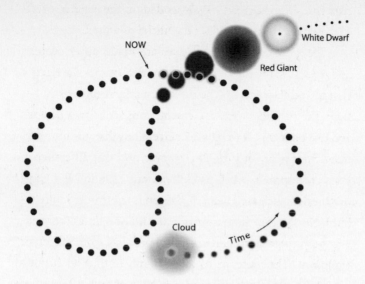

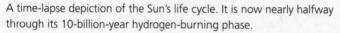

1.1
A time-lapse depiction of the Sun's life cycle. It is now nearly halfway through its 10-billion-year hydrogen-burning phase.

Our conception of the Sun's life cycle is depicted in figure 1.1 in a time-lapse format, with 150 million years between successive frames. The proto-Sun condensed from a cloud of diffuse interstellar gas. Gravity pulled this entity together until its center was squeezed hot enough to trigger nuclear fusion of hydrogen into helium at a sufficient rate to balance the heat shining from its surface. Less than half of the Sun's central hydrogen has so far been used up, even though it is already 4.5 billion years old. It will keep shining for a further 5 billion years. It will then swell up to become a "red giant," large and bright enough to engulf the inner planets and to vaporize all life on Earth. During this red giant phase, lasting some 500 million years, hydrogen will continue to burn in a shell around the helium core. Next, the Sun will undergo a more rapid convulsion, triggered by the onset of helium fusion in its core. This action will blow off some outer layers—about a quarter of the Sun's mass altogether. The residue will become a white dwarf—a dense stellar cinder no larger than Earth—which will shine with a bluish glow, no brighter than today's full Moon, on whatever remains of the solar system.

To convey these vast timescales, an analogy can help. Imagine a walk across America, starting in New York when the Sun was born and ending in California when it is about to die. To pace yourself on this walk, you would have to take one step every 2000 years. All recorded history would be just a few steps. Moreover, those steps would be just before the halfway stage: somewhere in Kansas, perhaps—not the culmination of the journey. Likewise, even our Sun has more time ahead of it than has so far elapsed; and our entire universe could have an infinite future awaiting it.

We may still be near Darwin's "simple beginning": if life isn't prematurely snuffed out, our remote progeny will surely—in the eons that lie ahead—spread far beyond this planet. Even if life is now unique to Earth, there is time enough for it to "green" the entire Galaxy, and even to spread beyond.

Other Solar Systems

In the early twentieth century, our solar system was suspected of being the outcome of a close encounter between our Sun and a passing star, which tore from it a stream of gas. This stream condensed into droplets, each a protoplanet. Stars are very thinly spread through space, however: in a scale model where the Sun was the size of a tennis ball, the nearest stars would be thousands of kilometers from the Sun and from one another. Close encounters would be freakishly rare; so also, if this theory were right, would be planetary systems. But such "catastrophist" views fell from favor in the second half of the century. Astronomers came to prefer an alternative theory that rendered planets a natural concomitant of star formation.

When interstellar gas contracts to form a star, its density rises a billion billion times. Any slight spin in the original gas would have been so much amplified during the collapse (a cosmic version of what happens when pirouetting ice skaters pull in their arms) that centrifugal forces would prevent all of it from contracting to the size of a star. Instead, any protostar, as it contracted, would naturally spin off a disk. In these disks, made from the gas and dust that pervades interstellar space, dust particles would stick together to make rocky "planetesmals," which in turn merge to make planets.

Disks have now been detected around newly forming protostars in the Orion Nebula and elsewhere. (Their discoverers coined the name "proplyds," short for protoplanetary disks). Proplyds are a natural concomitant of any star's birth, so there is every reason to expect other stars to be orbited by retinues of planets.

Even if they were orbiting one of the nearest stars, planets would be too faint to be seen directly with present-day telescopes. A planet would appear fainter than its parent star by a huge factor—roughly the same factor, in fact, by which Venus and Jupiter appear to us fainter than our Sun. But in the last few years, planets have been revealed indirectly through their effect on their parent star. Some stars have been found to be wobbling slightly in their positions, just as would be expected if planets were orbiting around them. A planet tugs the star around in a small counterorbit, rather like a small dog pulling its owner on a leash.

The first success was achieved by two Swiss astronomers, Michel Mayor and Didier Queloz. They analyzed the light from a nearby star, 51 Persei, which closely resembles the Sun, to see if there were slight changes in its color (or, in physical terms, in the wavelength of its light). If an object moves toward us, the waves get bunched up and appear to have shorter wavelengths—in other words, they shift toward the blue end of the spectrum. Conversely, light seems reddened if its source recedes from us. This is analogous to the so-called Doppler effect for sound waves, whereby the pitch is higher if the source approaches us. (Doppler famously used trumpeters on a railway truck to demonstrate this now-familiar effect.) Mayor and Queloz found that the light from 51 Persei shifted slightly toward the blue, and then toward the red, then toward the blue again; the pattern repeated regularly. This regularity implied that the star had a near-circular motion, induced by an unseen planet: star and planet pivot around the system's center of mass, so that the presence of the planet makes the star itself move. The inferred planet is about the size of Jupiter and orbits at around 50 kilometers per second. The star itself moves at only 50 meters per second— one thousand times slower than the planet because it is one thousand times heavier.

It was a real technical triumph to detect these slight motions. Geoffrey Marcy and Paul Butler in California are now the champion planet hunters, having used the same technique as Mayor and Queloz to find planets around dozens of other stars. This technique measures just the part of the star's motion that is directed along our line of sight—a small *transverse* motion does not show up as a Doppler shift. But specially instrumented telescopes should soon be able to reveal the tiny side-to-side motion induced by an orbiting planet by detecting slight changes in a star's position on the sky. In 1999, Marcy and Butler discovered that the nearby star Upsilon Andromedae had at least three Jupiter-sized planets: one in a very close circular orbit, with period 4.6 days; the other two in larger, slower orbits. More and more stars are now being revealed to have orbiting planets. The systems display surprising variety: there are planets that are up to about twenty times as heavy as Jupiter; the orbital periods can be as short as a few days; and the orbits are sometime near-circular, but surprisingly often they are highly elliptical.

The eventual goal, of course, is to have a sharp and sensitive image that reveals orbiting planets directly. This kind of image is still something for the future. But some claimed planets have already revealed themselves in other ways—for instance, by causing slight changes in the apparent brightness of their parent star. If a planet moved across the face of the star (as, in our own solar system, the planet Venus occasionally transits the face of the Sun), then the star would dim slightly each time, once per orbit, that the planet passed in front of it. This technique only works, of course, if our line of sight is close to the plane of the orbit.

Other Earths?

The planets found so far, orbiting solar-type stars, are all roughly the size of Jupiter or Saturn. But there is every reason to suspect that these are the largest planets in other "solar systems" whose smaller planets are not yet detectable. Planets the size of Earth would induce motions of merely centimeters per second, not meters, in the central star—too small to be discerned by current techniques. They are also hard to find by any other method. If such a planet were to move in front of a star, it would reduce its brightness by less than one part in 10,000. The best hope of detecting this minuscule dimming would be to use a telescope in space, where the starlight is unaffected by Earth's atmosphere and therefore is steadier. A planned European space mission called "Eddington" (named after the famous English astronomer) should be able to detect transits of Earthlike planets across bright stars within the next decade. The longer-term goal is to observe Earth-sized planets directly, rather than just inferring them. This capability will require very large telescope arrays in space—and it is far from a crazy idea.

From my home base in England, I watch the U.S. space program with interest and admiration. It is far larger than Europe's, its scale being a legacy of superpower rivalry. I am underwhelmed by the International Space Station. But it is better news that NASA's somewhat messianic chief executive, Dan Goldin, has focused the less costly unmanned program on the scientific theme of "Origins," and has included the so-called Terrestrial Planet Finder, capable of detecting planets as small as our Earth, as a main thrust of that program. In Europe, a similar project, called "Darwin," is also being planned.[2]

We were all, when young, taught the layout of our own solar system—the sizes of the nine major planets and how they move in orbit around the Sun. But, twenty years from now we shall be able to tell our grandchildren far more interesting things on a starry night. Nearby stars will no longer just be points of light—we will think of them as the Suns of other solar systems. We will know the orbits of each star's retinue of planets, and the sizes (and even some topographic details) of the bigger ones.

We will be especially interested in possible twins of our Earth—planets the same size as ours, orbiting other Sunlike stars, having temperate climates. We still don't know how many of these objects there are.

Most of the systems so far discovered, incidentally, are surprisingly different from our own solar system and offer rather poor prospects for habitable planets. Many contain Jupiterlike planets on eccentric orbits much closer to their parent star than our own Jupiter is. These massive bodies would destabilize any Earthlike planet in a near-circular orbit at the "right" distance for its parent star. This discouraging finding may be partly the outcome of observational selection: fast-moving heavy planets, orbiting close to their parent star, have been the easiest for Marcy and Butler to detect. We cannot yet be sure what fraction of planetary systems would permit an Earthlike planet to survive undisturbed for billions of years in a near-circular orbit; but among the many millions of planetary systems (formed with one, two, or three high-mass planets), there would surely be some planets on Earthlike orbits, with temperatures such that water neither boils nor stays frozen.

2 Life and Intelligence

Life's Likelihood

An iconic image from the 1960s was the first photograph of the entire round Earth, taken from our Moon. Our habitat of land, oceans, and clouds was revealed as a thin, delicate-seeming glaze. Our home planet—the "third rock from the Sun"—is very special. The beauty and vulnerability of "spaceship Earth" contrasts with the stark and sterile moon-scape on which the astronauts left their footprints. It may take more than twenty years before we can hang on our walls a poster of another Earth, but when we can, it will surely have even more impact than the classic picture of our home planet.

But even before we are able to image such planets in detail, we will have learned much about them from the Terrestrial Planet Finder space-telescope array or from its European counterpart. Viewed from, say, ten light-years away—the distance of a nearby star—Earth would be, in Carl Sagan's phrase, a "pale blue dot," seeming very close to a star (our Sun) that outshines it by a factor of many billions. The shade of blue would be slightly different, depending on whether the Pacific Ocean or the Eurasian land mass were facing us, and the brightness would vary seasonally. By observing other Earths, even if we can't resolve any detail on their surfaces, we can infer the length of their day, their climate, and even their gross topography.

From the spectrum of its light and infrared radiation, we could infer what gases existed in such a planet's atmosphere. Detection of ozone, implying an atmosphere rich in oxygen, would strongly indicate a biosphere. Our own atmosphere did not start out oxygen rich but was transformed thus by primitive bacteria in its early history. Spectroscopic evidence of atmospheric methane—the product of decaying vegetation or flatulent ruminants—could be another signature of an Earthlike biosphere.

How likely is it that other worlds harbor life? That's the question that galvanizes NASA and the wider public. In a propitious environment, what is the chance that simple organisms would emerge? Does life start on any planet in the right temperature range, where there is water, along with other elements such as carbon? Could some of these planets, orbiting other stars, harbor life forms far more interesting and exotic than anything we might find on Mars? What is the chance that some forms have evolved into something that can be called intelligent?

So long as we know of only one biosphere, we cannot assess whether it is a likely or an unlikely phenomenon. Life may be ubiquitous in the cosmos, or it may have huge odds stacked against it. Explorations of the solar system in the coming decades may firm up the odds. Mars is still the main focus of attention. Since the 1960s, space probes have revealed dramatic Mars-scapes: volcanoes up to 20 kilometers high, and a canyon 6 kilometers deep and stretching for 4000 kilometers. But most remarkable of all are the dried-up river beds — evidence that Mars may once have been hospitable to life, even if this is not the case today. There are even stratified features that look like the shoreline of an ocean. If there was indeed, sometime in the past, a large area of surface water, it is likely to have originated deep under ground and was forced up through thick permafrost. Mars was therefore never a location for prime beach-front property.

Back in the 1970s, NASA made its first serious exploration of the Martian surface. The Viking probe landed onto a barren rock-strewn desert, and its robot arm scooped up samples of soil, which were analyzed by onboard instruments. No firm sign of even the most primitive life was evident. The only serious claim for fossil life on Mars came later and was based on analysis of a piece of Mars that made its own way to Earth. Mars is continuously battered by impacts that throw debris out into space. Some of the resulting rocks, after wandering in orbit for many millions of years, crash down on Earth as meteorites. In 1996, NASA officials announced that a meteorite recovered from the Antarctic carried what might be traces of Martian life. At an elaborately orchestrated press conference, President Bill Clinton pronounced: "`Today, Rock 84001 speaks to us across these billions of years and millions of miles. It speaks of the possibility of life." The evidence was of two kinds: complex organic molecules similar to those found on Earth when organisms decay; and what looked like fossils of bacteria-like creatures, though even smaller than actual bacteria. Despite the hype, the strength of these claims is still disputed: life on Mars may vanish just as the "canals" did a century ago.

But hope has not been abandoned. Over the next few years, an armada of space probes will be launched toward the Red Planet to analyze its surface, to fly over it, and, in later missions, to return samples to Earth. In 2004, the European Space Agency's "Huygens" probe, part of the cargo of NASA's Cassini mission to Saturn, will parachute into Titan's atmosphere, seeking anything that might be alive on this giant moon. Other possible sites for life are Jupiter's frozen moons, Europa and Ganymede: there are plans to land submersible probes that could seek life beneath their ice-covered oceans.

So long as we have knowledge only about our own biosphere, we can't rule out the possibility that life on Earth is the outcome of an extraordinary chain of events, one so unlikely that it does not occur around any of the other 10^{21} stars within range of our telescopes. We still know too little about how life began to dismiss this possibility. But suppose life had emerged twice within our own solar system. Then it couldn't be a fluke: life would surely be common on planets around other stars as well. Therefore it is important to detect life on one of the other planets or moons of our solar system. There is an important proviso, however: before drawing any inference about the ubiquity of life, we would need to be quite sure that any nonterrestrial life had indeed begun independently, and that organisms had not made their way, via cosmic dust or meteorites, from one planet to another.

Bruno's Dream: Alien Intelligence?

Even if there is life elsewhere within our solar system, nobody expects that it would be anything but primitive. Our galaxy contains millions—perhaps billions—of other planets orbiting other stars. What are the prospects that there are advanced life forms on some of these bodies?

In 1584 Giordiano Bruno, a Dominican monk, published *On the Infinite Universe and Worlds.* Bruno had, a decade earlier, fled from his Naples monastery and enjoyed a peripatetic European existence. But in 1592 he imprudently returned to Italy—lured, it is thought, by the hope of a professorship at Padua which instead went to the young Galileo—and fell into the clutches of the Inquisition. He was imprisoned in Rome for his "obstinate and pertinacious heresies." In February 1600, he was burned at the stake in the Campo de Fiori, Rome, where he is now commemorated by a fine bronze statue.

Among Bruno's conjectures was the following: "There are countless constellations, suns and planets; we see only the suns because they give light; the planets remain invisible, for they are small and dark. There are also numberless earths circling around their suns, no worse and no less than this globe of ours." In the last years of the twentieth century, his prescient belief was vindicated: there are, assuredly, planetary systems around many other stars

Bruno had a further conviction: "No reasonable mind can assume that heavenly bodies which may be far more magnificent than ours would not bear upon them creatures similar or even superior to those upon our human earth." This idea was, of course, then a fantasy, but the concept of a "plurality of inhabited worlds" has, ever since, had surprisingly sustained support. The great eighteenth century astronomer William Herschel, discoverer of the planet Uranus, even thought that the Sun was inhabited; and a hundred years ago, many believed Martians existed. Even though our conception of the physical universe has been transformed since Bruno's time, we still cannot gauge the likelihood of extraterrestrial intelligence.

Despite this enduring ignorance—or maybe because of it—discussion of this subject is polarized. Some side with Bruno; others argue dogmatically that we are alone. For myself, I think agnosticism is the only rational stance on this issue. We don't know enough about life's origins—still less about what natural selection can and cannot do—to say whether intelligent aliens are likely or unlikely.

To firm up the odds on finding extraterrestrial intelligence, we need a clearer understanding of just how special Earth's physical environment had to be in order to permit the prolonged selection process that led to human life. Some astronomers claim that Earth is exceedingly special and that very few planets around other stars—even those that resembled Earth in their size and temperatures—would provide the requisite long-term stability for the prolonged evolution that must precede advanced life. Donald Brownlee and Robert Ward, in their book *Rare Earth*, give a dauntingly long catalog of prerequisites. The planet's orbit must not wander too close to its sun; nor too far away, as it would if other larger planets came too close and nudged it into a different orbit. Its spin must be stable (something that in our Earth's case is ensured by our large Moon). There must not be excessive bombardment by asteroids, requiring perhaps a Jupiter-sized planet in a nearly circular exterior orbit to act as a trap for stray asteroids. Furthermore, it must be orbiting a star with a specially propitious location in our Milky Way to ensure that it is neither unduly irradiated by cosmic rays, nor at high risk of encounters with other stars. If another star were to approach within the orbit of Jupiter, it would severely disrupt planetary motions, leaving Earth on a very different and perhaps highly eccentric orbit. But a star could endanger us, even if it did not come nearly as close as that: the Sun is surrounded by a huge reservoir of comets, mainly at safe distances far beyond the outer planets, but the gravitational pull of a star, even during quite a distant flyby, could deflect some of them into trajectories that crash onto Earth.

One can readily think of external environmental influences—frequent comet impacts, for example—that would impede biological evolution. But the prognosis for a biosphere is unknown even in the most benign physical environment. The most difficult and uncertain issues lie in the province of biology, not astronomy. There are two great questions, and it is important to distinguish them from each other. First, how did life begin? If we knew the answer, we would know whether life is a fluke or whether it is near-inevitable in the kind of primal "soup" expected on a young planet. But there is a second question: Even if simple life exists, what are the odds against its evolving into something that we would recognize as intelligent? This question is likely to prove far more intractable. Even if primitive life were common, the emergence of advanced life may not be.

We know, in outline, the key stages in life's development here on Earth. The simplest organisms seem to have emerged within 100 million years of the final cooling of the Earth's crust after the last major impact, some 4 billion years ago. But up to 2 billion years seem to have elapsed before the first eukaryotic (nucleated) cells appeared, and perhaps a further billion years before multicellular life emerged. The earliest evidence for most of the standard body types for animals dates from the so-called Cambrian explosion half a billion years ago. The immense variety of creatures on land emerged since that time—punctuated, of course, by major extinctions, such as the event 65 million years ago that wiped out the dinosaurs.

Simple life seems to have emerged quite quickly, whereas it took nearly 3 billion years for even the most basic multicellular organisms to come on the scene. This disparity of timescales suggests that there may be severe barriers to the emergence of any complex life. Intelligence could therefore be exceedingly rare even if simple life were widespread. Certainly our own emergence was the outcome of time and chance. If Earth's history were rerun, the fauna might be quite different. If, for instance, the dinosaurs hadn't been wiped out, the chain of mammalian evolution that led to humans may have been foreclosed. We can't say whether any other species would have taken our role.

Some evolutionists claim that even in a complex biosphere, the emergence of intelligent life was a fluke. Others, such as Simon Conway Morris, would dissent from this line. He notes that certain survival-promoting traits have led to the emergence of similar-looking animals via independent evolutionary tracks—for instance, many Australasian marsupials have close placental counterparts in other continents—and he argues that the same convergence might permit more than one evolutionary pathway toward intelligence. He writes: "My sense is that for all its immense diversity and plenitude, life is so constrained that what we see on Earth is far from being some sort of parochial backwater or provincial zoo, let alone a freak show. . . . There is a strong stamp of limitation, imparting not only a predictability to what we see on Earth, but by implication elsewhere."[1]

Exotic "Life"?

Searches for life will justifiably focus on Earthlike planets orbiting long-lived stars of the same type as our Sun. But science fiction authors remind us that there are more exotic alternatives. The physique of intelligent aliens would plainly depend on the habitat their home planet offered. For example, they could be balloonlike creatures floating in dense atmospheres; or, on a big planet where gravity pulled strongly, they could be the size of insects.

Perhaps life can flourish at lower temperatures, too, even

on a planet that has been flung into the frozen darkness of interstellar space—a planet whose main warmth comes from internal radioactivity, the process that heats the Earth's core.

Most recently-discovered planets are orbiting stars that resemble our Sun. But the very first extrasolar planets to be discovered are in a more exotic environment. In 1992 the radio astronomer Alex Wolszczan discovered three planets, each smaller than Earth, orbiting not an ordinary star but a neutron star—an object so dense that a mass larger than the Sun's is squeezed within a radius of 10 kilometers, with a density 100,000,000,000,000 times higher than an ordinary solid.[2]

Neutron stars are dense cinders left behind after supernova explosions. They interest physicists because they exemplify extreme conditions that could not possibly be simulated in a terrestrial laboratory and thereby offer the chance to test natural laws to their breaking point, and perhaps to learn something fundamentally new. The radiation that they emit is vastly brighter than a terrestrial laser, and the magnetic fields on the surface are millions of times more intense than can be created in laboratories. The force of gravity on a neutron star is 1000 billion times stronger than on Earth: when we drop a pen from table height, it makes a noise; the same event, on a neutron star, would release as much energy as a kiloton of high explosives.

The best-known neutron star lies in the center of the Crab Nebula, the expanding debris from a supernova witnessed by Chinese astronomers in A.D. 1054. The chief astrologer and computer of the calendar, Yang Wei Te, recorded it as a "guest star" and deemed it "yellow and favorable for the

Emperor." The same supernova was recorded in Korea and Japan; it is even claimed that it features in some Native American drawings. The neutron star in the Crab Nebula spins at thirty-three revolutions per second and emits a beam of intense radiation that we can observe once per revolution each time it sweeps, like the beam of a lighthouse, across our line of sight. Many hundreds of other such "pulsars" are known. Most are the remnants of supernovae that occurred millions of years ago. The debris from these earlier explosions would no longer be visible like the Crab Nebula, but it would have long ago dispersed and mixed into diffuse interstellar gas.[3]

The neutron star observed by Wolszczan radiates very little light, though it emits penetrating X-rays and ejects a wind of particles at nearly the speed of light. It does not provide an environment propitious for life. Indeed, it is hard to understand how this particular system of planets got there. It could have formed after the supernova from gas that rained down toward the neutron star, forming a disk, but a more likely scenario is that the precursor star already had its retinue of planets and then became a supernova. There could have been enough time for the planets to have developed simple life forms if some of them were on orbits farther out than the three that have been detected. But such life would not have survived the event that formed the neutron star. Life may have emerged, only to be snuffed out in the vaporizing blast from the supernova.

It doesn't take an explosion to snuff out life. Our Sun will never become a supernova and will in time leave behind a white dwarf rather than a neutron star. But one wonders about the prospects for any intelligent life on a planet whose

LIFE AND INTELLIGENCE

central Sunlike star became a giant and blew off its outer layers. Such considerations remind us of the transience of inhabited worlds, and that any seemingly artificial signal that we might conceivably detect sometime could come from superintelligent (though not necessarily conscious) computers, created by a race of alien beings that had already died out.

Perhaps life could emerge on a neutron star itself—a scenario described in a science fiction classic by Robert Forward. The kinds of hyperdense microscopic organisms that could exist in this environment, controlled by nuclear forces, would have a metabolism far faster than ordinary chemically based life, so an entire ecosystem could emerge and evolve in a few years.

At the other extreme, authors of fiction have reminded us that there could be diffuse living structures, freely floating in interstellar clouds. They would live (and, if intelligent, think) in slow motion, but nonetheless they might come into their own in the long-range future (see chapter 7).

Interstellar Discourse

Claims that advanced life is widespread must confront the famous question first posed by the great physicist Enrico Fermi: "Why aren't the aliens here?" Why haven't they visited Earth already, or at least manifested their existence in a way that leaves no doubt? Why aren't they, or their artefacts, staring us in the face? This argument gains further weight when we realize that some stars are billions of years older than our Sun: if life were common, its emergence should have had a head start on planets around these ancient stars.[4]

Even if we have not been visited (and of course we cannot be absolutely sure we haven't) we should not, despite Fermi's question, conclude that aliens don't exist. It would be far easier to send a radio or laser signal than to traverse the mind-boggling distances of interstellar space. Indeed, we can already send signals that could be picked up by a civilization with the same technology as ours that was in orbit around a nearby star. But even the nearest stars are so far away that signals would be in transit many years. (Aliens equipped with large radio antennas could, in any case, pick up the strong signals from antiballistic missile radars, as well as the combined output of all our TV transmitters. If they could decode them, it's not hard to speculate what they might conclude about "intelligent" life on Earth.)

It makes sense first to listen rather than to transmit. If a signal were detected, there would be time to send a measured response, but there is no scope for snappy repartee: any two way exchange would take decades at best. It is of course possible to send coded pictures, or even blueprints of three-dimensional structures—either of artifacts or molecular templates such as a genome. In the long run, a dialogue could develop. The logician Hans Freudental proposed an entire language, called "lincos," for interstellar communication, showing how it could start with the limited vocabulary needed for simple mathematical statements, and then gradually diversify the realm of discourse.

Searches for Extraterrestrial Intelligence (SETI) are a worthwhile gamble, even if one suspects that there are heavy odds against success, because of the huge philosophical import of any detection. A manifestly artificial signal—even if it was as boring as a set of prime numbers or the digits of "pi" in binary notation—would convey the momentous message that intelligence (though not necessarily consciousness) is not unique to Earth and had evolved elsewhere, and that concepts of logic and physics are not peculiar to the kind of hardware we carry around in our heads.

The SETI Institute at Mountain View, California, is spearheading these searches; its work is supported by hefty private benefactions. Any interested amateur with a home computer can download and analyze a short stretch of the data stream from a radio telescope. At the time of writing, three million people had taken up this offer—each, no doubt, inspired by the hope of being the first to find ET. In the light of this broad public interest, SETI searches seem to have a surprisingly hard time getting public funding, even at the level of the tax revenues from a single science-fiction movie. If I were an American scientist testifying to Congress, I would be happier requesting a few million dollars for SETI than seeking funds for more specialized science, or indeed for conventional space projects.

Even if intelligence were widespread, we may never become aware of more than a small and atypical fraction of what is out there. Some brains may package reality in a fashion that we can't conceive. Others could be uncommunicative: living contemplative lives, perhaps deep under some planetary ocean, doing nothing to reveal their presence. There may be a lot more life out there than we could ever detect. Absence of evidence is not evidence of absence. The only type of intelligence we could detect would be one that led to a technology we could recognize.

Other "Suns," many light-years away, may shine on alien habitats as intricate and complex as our own. All would have distinctive evolutionary histories. Not even the most extremely convergent view of evolution would lead us to expect that other biospheres in our Galaxy resemble ours in detail. There is less chance of two ecologies repeating themselves than of two monkeys typing out the same Shakespearean play. It could indeed happen if space and time were literally infinite, but not within the compass of the vast, but finite, domain that we can actually observe. All the monkeys in the world would have little chance of producing a single error-free sonnet, even if they typed for a billion years. And any ecosystem involves tremendously more variety than one language.

If the universe were literally infinite, then anything, however improbable, could happen. Indeed, it could happen infinitely often, leading to replicas of our Earth, even infinitely many of them. But these clones would be located far beyond our own Galaxy—indeed, far, far beyond the horizon of our observations.

It would in some ways be disappointing if searches for alien intelligence were doomed to fail. On the other hand, it would give humans a pretext for a boost in self-esteem: if our tiny Earth were a unique abode of intelligence (at least in the domain accessible to our telescopes), it would have greater cosmic significance than it would merit if the Galaxy already teemed with complex life. We would then have even stronger motives to cherish our pale blue dot in the cosmos and not foreclose life's future.

Our Sun has burned less than half its nuclear fuel and

will continue to shine for longer than it has taken our biosphere to evolve from simple beginnings on the young Earth—several times longer, in fact, than the time since the first multicellular life form appeared. The broader cosmic future is far longer still—in later chapters I will suggest that it is infinite. Even if life is now unique to Earth, we should not conclude that our universe is not biophilic: life could still prevail, eventually becoming pervasive and even taking over in the cosmos. There is plenty of time for life seeded from Earth to spread through the entire Galaxy, and even beyond. In this perspective, we are not of course envisaging anything recognizably human. Darwin himself wrote that, "Judging from the past, we may safely infer that not one living species will transmit its unaltered likeness to a distant futurity." And now, artificially genetic modifications can induce far faster changes than are possible through natural selection.

The spread of life from Earth could happen provided human actions do not foreclose this long-range prospect. Earth is perpetually at risk from encountering an asteroid large enough to cause worldwide devastation—ocean waves hundreds of meters high, tremendous earthquakes, and changes in global weather. Movies like *Deep Impact* have boosted public awareness of such potential catastrophes, even though they may have been alarmist in blurring the fact that the risk, in the lifetime of anyone alive today, is less than one in 10,000. (This chance is, however, no lower than the risk for the average person of being killed in an air crash. Indeed, it is higher than any other natural hazards that most Europeans or North Americans are exposed to, and for that reason it seems fully worthwhile to devote modest efforts to survey the sky for potentially dangerous Earth-crossing asteroids.)

We also run a less quantifiable risk—though, one fears, a larger and sharply increasing one—of being wiped out by some catastrophe that we bring upon ourselves, perhaps some experimental misadventure or a terrorist act that deploys techniques from bioscience. However, once there were self-sustaining communities away from Earth, humans would, as a species, be invulnerable to any such catastrophe. This line of thought offers, in my view, the strongest motive for pursuing a program of manned space flight, even though advances in robotics and miniaturization are weakening the practical case for it. Development of space habitats—something that will be feasible before the end of the twenty-first century—would provide insurance against the risk that our species could be extinguished and its potential foreclosed.

Our Own Outward Odyssey?

In July 1969, Neil Armstrong's "one small step" on the Moon made space travel a reality. At the time, this step seemed just a beginning. Most of us then imagined follow-up projects: a permanent lunar base, rather like the one at the South Pole; or even huge space hotels that orbit Earth. Manned expeditions to Mars seemed the natural next step after that. But none of these possibilities has yet happened. The year 2001 does not resemble Arthur C. Clarke's depiction any more than 1984 (fortunately) resembled Orwell's.

The program, announced by President John F. Kennedy

in 1961, "to land a man on the Moon before the end of the decade, and return him safely to Earth," was lavishly funded for reasons of superpower rivalry. Reaching the Moon was an end in itself: the impetus fizzled out, and no lunar expedition has taken place since 1972.

How long will it be before people return to the Moon, and perhaps explore still farther afield? The centerpiece of the U.S. program is the new International Space Station: this massive structure, the size of a football field, will be in permanent orbit several hundred kilometers up. Even if it is finished—something that seems uncertain, given the immense and ever-rising costs and prolonged delays—it will seem uninspiring if it merely enables people to circle round and round the Earth at a very high cost. In itself this will not seem a very exciting feat, thirty years after men walked on the Moon. For most of science, with the obvious exception of space medicine, and certainly for most astronomy, the Space Station is as suboptimal a base as an ocean liner would be for ground-based telescopes. Its main scientific use will be as an assembly base for large and fragile equipment (for instance, gossamer thin mirrors for huge space telescopes), which could then be gently tugged into a more remote orbit.

The Space Station could in time also serve as a staging post on the way to other planets. But such adventures will have to await some technical breakthrough that renders space travel much cheaper and easier. Present launching techniques are as extravagant as air travel would be if the plane had to be overhauled after every flight. Space flight will only be affordable when it adopts the same techniques as supersonic aircraft. It could also be a lot cheaper if it were undertaken as a dangerous enterprise by crews with the attitude of test pilots or round-the-world yacht racers. In the long run, space exploration may not be a government project, but privately funded—perhaps, indeed, becoming the province of wealthy adventurers prepared to accept high risks to boldly explore this far frontier.

Space travelers, wherever they go, will confront a hostile environment. It would be a daunting task to build a comfortable habitat, either free-floating in space or on the Moon or Mars. But it could be done: the necessary infrastructure could be set up in advance, by unmanned probes and robots. Raw materials could be mined and processed on-site, and it may eventually be possible for communities to live independently of supplies from Earth. When self-sustaining communities have been established away from Earth—on the Moon, on Mars, or freely floating in space—our species will be invulnerable to any global disaster, and whatever potential it has for the 5-billion-year future could not be snuffed out.

3 Atoms, Stars, and Galaxies

Starstuff

The French philosopher Auguste Comte averred in 1835 that, while we might learn the sizes and motions of stars, we would never know what they were made of. Within just twenty years, this pessimism proved misplaced.

Newton's famous experiment with a prism had shown that sunlight could be split into a spectrum, displaying all the colors of the rainbow. Early in the nineteenth century, the German optician Josef von Fraunhofer used a more elaborate instrument to display the Sun's spectrum as a ribbon of light that he could then analyze with a microscope. He found that it was crossed by many dark bands, indicating that certain particular shades of color (corresponding to light of specific wavelengths) were missing. Robert Bunsen (of Bunsen burner fame) and Gustav Kirchoff analyzed the light emitted by glowing gases in the laboratory and discovered similar patterns of spectral lines. Each kind of atom emits light with a specific set of precise colors—for instance, the yellow light due to sodium, or the blue glow, familiar from street lamps, due to mercury vapor. If the atoms are in front of a hotter light source, then they absorb these same colors, and yield a spectrum with a distinctive pattern, similar to a modern-day bar code. The dark lines on the solar spectrum are due to absorption by a somewhat cooler gas that overlies the Sun's blazing surface.

The advent of photography revolutionized astronomy: light could accumulate over a long exposure, revealing features too faint to be seen just by looking through a telescope. A wealthy English amateur astronomer, William Huggins, used this newly available technique to study the light from stars; he found that, when starlight was spread out into a spectrum, it displayed the same line patterns that had already been seen in the solar spectrum. The Sun was made of the same stuff as Earth; and so were the stars.

Stellar spectra reveal the bar codes of many different elements. But there is no straightforward relation between the strength of a spectral feature and the abundance of the element that causes it. The prominence of a particular element's signature depends, we now realize, on details of atomic physics and on the temperature and structure of a star's outer layer. These issues were not settled until well into the twentieth century.

Indeed, until the 1920s it was not appreciated that hydrogen and helium—the two simplest elements—were overwhelmingly the most abundant in our solar system and elsewhere. Because of their volatility, these very lightweight atoms are underrepresented on Earth, but they amount to 98 percent of the Sun's mass and the mass of the giant outer planets. Much credit here goes to Cecilia Payne, whose 1925 Ph.D. thesis was pivotal. Sadly, the skepticism of the most eminent expert of the time, Princeton's Henry Norris Russell, pressured her to downplay her greatest discovery, and to caution her readers that the inferred abundance of hydrogen and helium "is improbably high and almost certainly not real." Payne had a distinguished career and received belated recognition as a Harvard professor. (Unfortunately, this was by no means the last time that outstanding contributions by women scientists went underrecognized.)

Stellar Alchemy

It gradually became clear that the relative abundances of different elements displayed regularities from star to star, and that these proportions resembled those found in our solar system (fig. 3.1). How did this particular mix arise? By the 1930s, both stellar and nuclear physics were well enough developed to address this question. Could the elements be the outcome of nuclear transmutations in stars?

Astronomers can compute the life cycle of stars that are, say one-tenth, ten, or thirty times as heavy as the Sun.

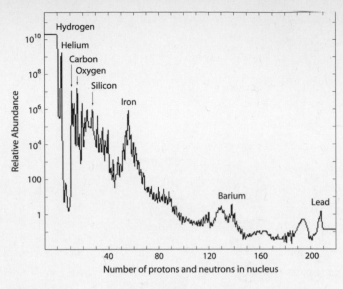

3.1
Abundances of elements.

Figure 3.2 shows some time-lapse depictions of how these different types of stars evolve: the heavier stars are far brighter and have shorter lives because they exhaust their fuel more quickly. Even the heaviest stars, however, live so long compared to astronomers that we have in effect a single snapshot of each star's life. But we can check our calculations, in principle, much as a newly landed alien could quickly infer the life cycle of trees or people.

We can observe entire populations of stars. The best test beds are star clusters—swarms of up to 100,000 stars of different sizes that formed at the same time. Indeed, there are places where stars still seem to be forming: for instance, in the pillars of the Eagle Nebula. These remarkable gaseous structures, about seven thousand light-years away, feature in one of

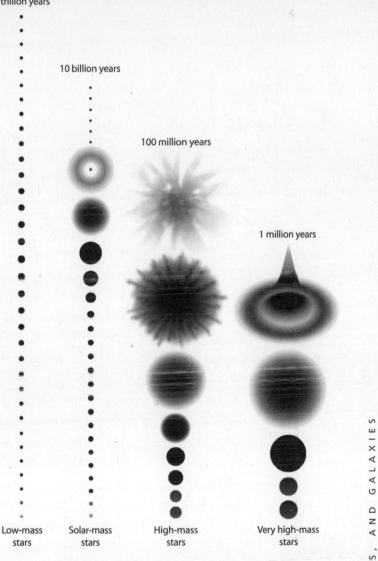

1 trillion years

10 billion years

100 million years

1 million years

Low-mass stars   Solar-mass stars   High-mass stars   Very high-mass stars

3.2
Time-lapse depictions of life cycles of stars with different masses (with time increasing upwards).

the most widely reproduced of all Hubble Space Telescope images. Cool interstellar clouds harbor bright young stars as well as newly condensing protostars that are orbited by dusty disks from which planets may form.

The changes in stars are generally imperceptibly slow on human timescales, but not always. Massive stars end their lives violently by exploding as supernovae. The nearest supernova observed in the twentieth century was detected in 1987: a bright object appeared in the sky on 24 February that had not been visible the previous night. After a few weeks, it started to fade, and its development is still being monitored, using all the techniques of modern astronomy. It gives theorists a chance to check their elaborate computer simulations of supernova explosions and exemplifies how the cosmos offers us the chance to study "extreme physics"—to check the laws of nature under conditions than can not be reproduced in laboratory experiments.

In one thousand years, the ejecta from the 1987 supernova will look like the famous Crab Nebula, the blue-glowing remnant of a supernova (already mentioned in chapter 2) that was detected in A.D. 1054. After 10,000 more years, the expanding debris would have faded and merged with the ambient interstellar gas.

Supernovae fascinate astronomers, but only one person in 10,000 is an astronomer. What is the relevance of these explosions, thousands of light-years away, to the other 9999? The answer is that they hold the key to an everyday terrestrial question: Where did the atoms we are made of come from, and why are some common—oxygen, carbon, and iron, for instance—while others, like gold and uranium, are very rare?

Stars more than ten times heavier than the Sun consume their central hydrogen hundreds of times more quickly than the Sun does, and consequently they shine much brighter. Gravity then squeezes them further, and the centers get still hotter, until helium nuclei can themselves stick together to make the nuclei of atoms further up the periodic table. Such stars develop an onion-skin structure: a layer of carbon surrounds a layer of oxygen, which in turn surrounds a layer of silicon; the core is mainly iron. Iron is the most tightly bound nucleus, and when a big star has turned its core into iron it faces an energy crisis. The core implodes, becoming a million billion times denser than an ordinary solid, turning into a neutron star or perhaps even a black hole. But this catastrophic collapse releases enough energy to trigger a colossal explosion—a supernova—that blows off the outer layers of the star. By the time this happens, a star contains the outcome of all the nuclear alchemy that kept it shining over its entire lifetime.

This account tells part of the story, but it cannot in itself account for all ninety-two naturally occurring elements. Iron is only number 26 in the periodic table; energy has to be added to it (rather than being released) by whatever processes form still heavier elements. These elements exist only in traces. Some of them—thorium and uranium, for instance—are forged in the heat of the supernova explosion itself. Others, like barium and bismuth, are built up by the capture of neutrons in red giant stars. Stars that are not heavy enough to become supernovae also make a contribution—indeed, they are the main source of carbon—because they blow off a wind of processed material as they burn.

The dominant figure in these insights, from the mid-1940s, was the Cambridge astrophysicist Fred Hoyle. He was a pioneer in understanding the internal structure of stars and how they evolve; he also had enough scientific background to understand the underlying nuclear processes. Still more important, he was able to enthuse other scientists, especially the nuclear physicist William Fowler, in whose laboratory at the California Institute of Technology many key experiments were done to elucidate the particular reactions that were critical. The key processes of stellar nucleogenesis were codified in a lengthy 1957 paper that Hoyle and Fowler coauthored with astronomers Geoffrey and Margaret Burbidge. (This classic article is referred to as "B^2 FH," the initials of its four authors).*

Our cosmic habitat is like an ecosystem. Gas is recycled through successive generations of stars (fig. 3.3). Fast-burning heavy stars transmute pristine hydrogen into carbon, oxygen, iron, and the rest of the elements in the periodic table; they then throw the spent fuel back into space, either via stellar winds or in the final supernova outburst. An oxygen atom expelled from a massive star may have wandered for hundreds of millions of years in interstellar space. It may then have found itself in a dense cloud, contracting under its own gravity to make a new star, surrounded by a dusty disk. That star might have been our Sun, and that particular atom could have ended up on Earth, perhaps some day to be cycled

---

*The Canadian physicist Alastair Cameron developed some key ideas independently. The details have been elaborated over subsequent decades, especially through a better understanding of what happens in the blast wave of the supernova explosion itself.

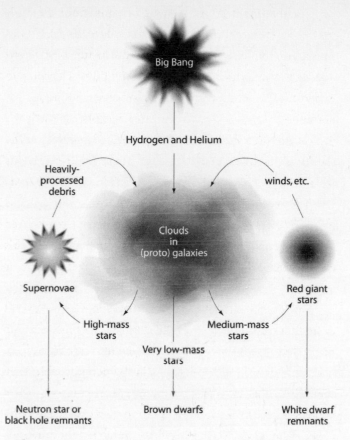

3.3

The Big Bang produces hydrogen and helium (plus a trace of lithium). This pristine gas condenses, under the action of gravity, into proto-galactic clouds. In these cloud, stars form. The smallest (and slowest-burning) stars are less than a tenth the Sun's mass; the heaviest are up to 100 times the Sun's mass. Nuclear fusion reactions in stars create heavier nuclei. Some of this "processed" material is expelled from medium and high-mass stars and later recycled into new stars. These processes can build up the elements of the periodic table in the proportions shown in figure 3.1. As galaxies get older, the heavy element abundance builds up. An increasing fraction of the original gas ends up as "dead" compact remnants.

through a human cell. To understand our origins, we must understand stars that formed long ago, in the remote parts of our Milky Way.

Aliens would be made of atoms similar to those we are composed of. And they would realize, as we have done, that these atoms themselves owe their existence to exploding stars. We and the aliens would have a shared affinity with the stars. We would all be stardust—or, more prosaically, the nuclear waste from the fuel that makes stars shine.

The network of nuclear reactions that forged the periodic table depends crucially on the microphysics: what happens when two atomic nuclei collide, which combinations of particles can stick stably together, and so forth. These reactions depend, in particular, on a balance between the two forces that control the main constituents within atomic nuclei: the electrical repulsion between protons, and the strong nuclear force between protons and neutrons. This nuclear force is now understood, on a somewhat deeper level, to depend on subnuclear particles called quarks and gluons.

Nuclear forces allow a periodic table of stable elements to exist and render nuclear power an efficient energy source in stars. Moreover, the synthesis of particular atoms turns out to be very sensitive to the strength of the nuclear force. For instance, each carbon nucleus (containing six protons and six neutrons) is made up from three nuclei of helium. As Fred Hoyle and the American astrophysicist Edwin Salpeter were the first to realize, this process works efficiently only if the carbon nucleus has a rather unexpected feature: a mode of vibration with a very specific energy. Otherwise, the chance of three helium nuclei combining together would be so low that carbon atoms—crucial, of course, for life—would be rare. Hoyle in fact suspected that carbon nuclei had this particular property before there was any experimental evidence for it. He cajoled his colleagues into doing a laboratory measurement, and the result vindicated his prediction. The vibration frequencies of atomic nuclei depend on the strength of the force that holds protons and neutrons together. Hoyle's coincidence would be lost if this force were changed by more than 1 or 2 percent, a fact that is often cited as evidence that our universe is governed by laws that are puzzlingly biophilic.

Other features of nuclei have important consequences. A neutron is heavier than a proton by 0.14 percent—little more than one part in 1000. But this difference, small though it is, exceeds the total mass of an electron. If electrons weren't so light, they would combine with protons to form neutrons, leaving no hydrogen. (In our actual universe, this combination happens only when it is induced by the extreme pressures inside neutron stars.)

The fact that electrons weigh very little compared to the nuclei of atoms is also important. This disparity is a prerequisite for the propensity of molecules like DNA to maintain their precise and distinctive structures. Heisenberg's uncertainty principle implies an inevitable fuzziness in the location of any particle, though the uncertainty is less for heavier particles. In a molecule, the uncertainty in an atom's position is determined by the mass of its nucleus. The orbits of electrons around the nucleus are very much larger, since electrons are lighter, and there is consequently more fuzziness in their positions. The electron orbits determine the overall size of the atoms and the spacing between the atoms in a molecule.

Because protons are 1836 times heavier than electrons, atoms can be quite precisely located relative to their neighbors; thus complex molecules can have well-defined shapes.

The marvelous story of stellar nucleogenesis disappointed George Gamow. As discussed in the next chapter, his belief that all elements emerged from the Big Bang (or "ylem," as he called it) was not borne out by detailed calculations. The temperature dropped so quickly during the first few minutes that there was no time for the network of necessary reactions to proceed; moreover, observations showed that younger stars contained more heavy elements than the oldest stars, corroborating the idea that stars formed from interstellar gas that became more "polluted" with time. But Gamow was at least partly vindicated; deuterium, helium, and lithium would have been made in the hot early universe. Even the oldest stars contain 23 percent helium—just the proportion that emerges from the Big Bang.

Stars, Atoms, and Gravity

Contemplative aliens, even if they were cloud-bound or eyeless, might discover enough about atoms and gravity to work out what stars are like. If we define stars as gravitationally bound fusion reactors, a very simple argument will allow us to work out how big they are—how many atoms we would have to pack together to make one.

Gravity is a very, very weak force. It pulls two protons together with a force about 10^{36} times weaker than the electric force between them. But in any large object, containing huge numbers of atoms, positive and negative electric charges almost cancel each other out: there are almost exactly as many electrons as protons. In contrast, everything has the same sign of gravitational charge, so the bigger an object is, the more important gravity is relative to other forces. Because gravity is so exceedingly weak—36 powers of 10 weaker than the electrical force between two protons—it triumphs in only very large objects indeed. Princeton physicist Robert Dicke was the first to emphasize that this is why stars are so big: the Sun, a typical star, contains about 10^{57} protons.* Dicke also estimated the time it takes for heat to diffuse out of a star, showing that, besides being big, stars have *long lives*, because gravity is so weak.

As a foretaste of speculations in later chapters, it's amusing to consider what the universe would be like if gravity weren't quite so weak. Suppose, for example, that gravity were "only" 26 rather than 36 powers of 10 weaker than the electric

---

*Dicke's argument follows from some fairly simple arithmetic. Suppose you assemble progressively larger lumps containing 10, 100, 1000 atoms, and so on. The 24th, containing 10^{24} atoms, would be the size of a sugar lump; the 40th would be the size of a mountain or a small asteroid.

The effect of gravity on each atom—how strongly gravity binds it to all the others in the lump—goes up in proportion to the total number of atoms but down by their average distance from each other. For each 1000-fold increase in mass, the importance of gravity goes up 100-fold. (This is because, though the number of atoms goes up by 1000, their average distance from each other goes up by 10.) Despite its initial handicap, amounting to 36 powers of 10, gravitational forces become dominant when more than about 10^{54} protons are packed together (36 being two-thirds of 54). This mass is about the same as that of Jupiter, the biggest planet in our solar system. To become a star, a body must be somewhat more massive still—containing more than 10^{56} atoms—so that it can hold itself together, gravitationally, even when its center is hot enough for nuclear fusion to occur. Figure A1 in the appendix depicts this argument graphically.

forces in atoms—but the microphysics were unchanged. Atoms and molecules would behave just as in our actual universe, but objects would not need to be so large before gravity became competitive with the other forces. In this imagined universe, stars would have 10^{-15} the Sun's mass. If these stars had planets around them, they would be smaller than the actual planets in our solar system by the same factor, but gravity on their surfaces would pull far more strongly than on Earth. Strong gravity would crush anything larger than an insect on hypothetical mini-planets around these mini-Suns. But more severe still is the limited time. Instead of living for 10 billion years, a mini-Sun would last for about one year and would have exhausted its energy before even the first steps in organic evolution had gotten under way.

In this hypothetical strong-gravity universe, the outlook for complex evolution would plainly be less propitious, because there would be less space and less time. There would be fewer powers of 10 between stellar timescales and the basic microphysical timescales for physical or chemical reactions. Although gravity is crucial in the cosmos, the weaker it is (provided it isn't zero) the grander and more complex can be its manifestations.

A universe that did not involve large numbers could not contain such a multilayered hierarchy of structures and would not allow time for complex evolution. A vast range of scales is a prerequisite for an "interesting" universe; this range comes about in our universe because of the enormous number 10^{36} that measures the weakness of gravity. In consequence of this weakness, cosmic objects controlled by gravity must be large and long lived.

4   Extragalactic Perspective

Other Galaxies

We live in a disk-shaped Galaxy that contains 100 billion stars. The Sun orbits around the central hub at a distance of 25,000 light-years, taking 200 million years to make a complete circuit (a galactic year). But our Galaxy is just one among many; for example, Andromeda, its nearest neighbor, lies about 2 million light-years away. To an observer on Andromeda, our Galaxy would look roughly as Andromeda looks to us—a vast, slowly swirling disk of stars and gas, seemingly oval shaped because it is viewed obliquely.

Galaxies are the basic ingredients of the large-scale universe, just as individual stars are the main ingredients of galaxies. There are about as many galaxies within range of our telescopes as there are stars in each galaxy. (Just as humans are about midway between atoms and stars, so galaxies are intermediate between stars and our observable universe.)

Stars are, basically, now well understood: we know enough about atoms and about gravity to calculate what a gravitationally confined fusion reactor would look like. The outcome of these calculations accounts for the sizes, brilliance, and colors of the stars we see and enables us to infer their ages and life cycle. Despite rapid recent progress, however, an equivalent understanding of galaxies still eludes us. Why is the large-scale cosmic scene dominated by these beautiful and distinctive aggregates of gas and stars? Why are some disklike, whereas others, the elliptical galaxies, are amorphous swarms of stars? The factors that stymie current attempts to answer this question will emerge later in this book: galaxies have a still-mysterious extra ingredient—dark matter—and some of their properties are imprinted right back at the beginning of our universe.

Galaxies are not sprinkled down randomly in space: most are in groups or clusters, held together by gravity. Our own Local Group, a few million light-years across, is dominated by two big galaxies, our own and Andromeda, but it also contains (at the last count) thirty-five small satellite galaxies, each held in a giant orbit by the gravitational pull of the entire Group. The Local Group is an outlying part of an archipelago of galaxies centred on the Virgo Cluster, whose core lies about 50 million light-years away. Such clusters are themselves organized in still larger aggregates. The nearest and most prominent of these giant features is the so-called Great Wall, a sheetlike array of galaxies about 200 million light-years away.

Large-Scale Structure and Expansion

If our universe contained clusters of clusters of clusters, ad infinitum—if it were what is called, in technical jargon, a simple fractal—then however large a volume we probed, the galaxies would still have a patchy distribution: we would simply be sampling larger and larger scales in the clustering hierarchy. But that is not how our universe looks (fig. 4.1).

Telescopes can now gather data far faster than ever before: by using optical fibers, astronomers can record hundreds of spectra at a time rather than having to observe galaxies one by one. Projects like the Sloan Digital Sky Survey, a five-year project to scan the sky with a specially instrumented telescope in New Mexico, are now systematically studying all galaxies out to about ten times the distance of earlier surveys. These deeper surveys reveal more clusters like Virgo and more features like the Great Wall. But there don't seem to be conspicuous structures on still larger scales. A box whose sides were 200 million light-years across would be capacious enough to contain a fair sample of our universe. Wherever it was placed, such a box would contain roughly the same number of galaxies, grouped in a statistically similar way into clusters, filamentary structures, and so on. And 200 million light-years is small compared to the horizon of our observations, which is about 10 billion light-years away.

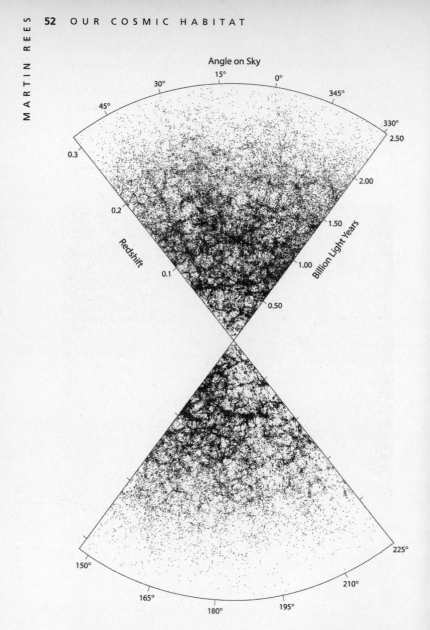

4.1
Large-scale clustering of galaxies: slices in the Northern and Southern Hemispheres as mapped with the Anglo-Australian Telescope.

We can well imagine a hypothetical universe that was indeed fractal-like (see fig. 4.2) and lacked the large-scale smoothness of our actual cosmic habitat. In such a universe, cosmology would be an even harder subject than it actually is. An analogy helps to explain why. Suppose you were on a ship in mid-ocean. If the sea were rough, you would see a complex

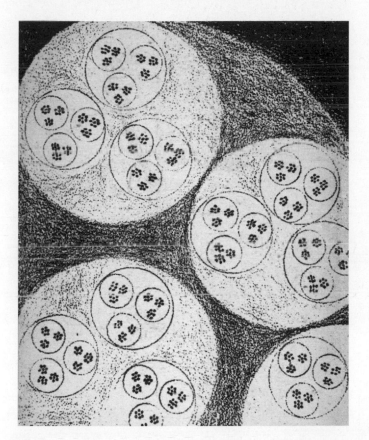

4.2
A depiction of a fractal-like universe, with clusters of clusters of clusters *ad infinitum*, as envisaged by the Swedish astronomer Charlier early in the twentieth century.

EXTRAGALACTIC PERSPECTIVE

pattern of waves, but the length, crest to crest, of even the longest ocean swells would still be small compared with the distance to the horizon; you could see a large enough sample of each kind of wave to gather statistics. On the other hand, in a mountain landscape (which can indeed be fractal-like) one peak can dominate the entire view. Cosmology is only tractable because our universe resembles an ocean surface rather than a mountain landscape. This large-scale smoothness allows us to define averages.

The overall motions are simple, too. Perhaps the most important broad-brush fact about our universe is that all galaxies (except for a few nearby objects in the same cluster as our own) are receding from us. Moreover, the redshift (a measure of the recession speed) is larger for the fainter and more distant galaxies. We seem to be in an expanding universe where clusters of galaxies are getting ever more widely separated—more thinly spread through space—as time goes on.

The simple relation between redshift and distance is called Hubble's law, after Edwin Hubble, who first claimed such a relation in 1929. It is conceptually better to attribute the redshift of a distant galaxy to a stretching of space while the light travels through the universe. The amount of redshifting—in other words, the amount that the wavelengths are stretched—tells us how much the universe has expanded while the light has been traveling toward us.

Hubble's law does not imply that we are in a special location. Imagine all the galaxies joined by rods. If these imaginary rods all stretched in the same proportion, so that any triangle made up of three rods kept the same shape as it expanded, then an observer on any galaxy would see the same pattern of expansion. It's as though space itself is expanding, carrying galaxies with it.*

Our universe may once have been squeezed to a single point, but everyone—whether on Earth, on Andromeda, or even on the galaxies remotest from us—can equally claim to have started from that point. We can not point to any particular location in our present universe and say, "That's the center": all galaxies are on the same footing.

Hubble could study galaxies that are only relatively close to us—those whose recession speeds were less than one percent of the speed of light. Thanks to larger telescopes, and (even more) to improved sensors for detecting faint light, the data now extend much farther, revealing galaxies so far away that their recession speeds are 90 percent the speed of light. The light from these very remote galaxies started to travel when they were several billion years younger and the universe was more compressed (in other words, when the imaginary rods joining them were all shorter). We must allow for the effects of relativity and the expansion of space in inferring the speeds and distances.

Telescopes: In Space and on the Ground

Progress in astronomy does not come from armchair theorists, even when they are aided by teraflop computers: it

---

*This relationship applies strictly to clusters rather than galaxies: in a cluster of galaxies, the individual members are influenced by the gravitational pull of the others, and there is no expansion. Nor is there any Hubble expansion *within* galaxies.

mainly depends on observations and on high-precision instruments right at the limit of technology. Astronomy has a longer history than any other science, except perhaps medicine. It was the first to involve precise measurement, and perhaps the first to do more good than harm. It was certainly the first to need large instruments. Astronomical artifacts—astrolabes, clocks, and telescopes—have been (and still are) monuments to human ingenuity and, as a bonus, often of great aesthetic appeal as well.

Throughout the last century, optical telescopes were built with progressively larger mirrors, enabling them to collect more light. By the 1980s, more than a dozen had mirror diameters exceeding 4 meters. The power of these telescopes dramatically improved when traditional photographic film was superseded by light-sensitive microelectronic chips called charged-coupled devices. The efficiency for detecting faint light went up from 1 to 80 percent of the theoretical maximum, giving these 4-meter telescopes an immense boost. But, having achieved such high efficiency, the only way to detect still fainter objects is by collecting yet more light, and this requires even larger mirrors. Telescopes with mirrors 8 or 10 meters across can collect four times more light from very faint and remote galaxies than earlier-generation ground-based telescopes. The two Keck Telescopes in Hawaii are now being complemented by several more. Most impressive of all is the unimaginatively named Very Large Telescope, a connected cluster of four telescopes, each with an 8.2-meter mirror, constructed in the Chilean Andes by a consortium of European nations.

The Hubble Space Telescope project was dogged by delays, cost overruns, and technical problems, but it eventually fulfilled the astronomers' hopes. Its out-of-focus mirror was corrected by the first manned repair mission in 1994; and the light detectors on board have been upgraded by later repair missions. Barring mishap, it could continue functioning until 2010. Although its mirror diameter is only 2.4 meters, its location above the blurring effects of Earth's atmosphere has enabled it to produce the sharpest and deepest pictures yet of very distant regions.

Up in space, the Hubble Telescope will be followed, a few years from now, by the Next Generation Space Telescope, an international project planned to have a mirror 6 or even 8 meters across and to be sensitive to infrared light. This instrument will probe still farther into the remoteness of space—and perhaps back in time toward a dark age before any real galaxies formed, when the only sources of light would have been small "subgalaxies."

Down on the ground, 10-meter telescopes will be supplemented by a new generation of still more ambitious "giant eyes." There are serious plans in Europe to develop the so-called OWL (OverWhelmingly Large) Telescope, which would have a 100-meter mirror—actually a mosaic of many less gigantic elements that could be individually adjusted to compensate for fluctuations in the atmosphere. This futuristic ground-based instrument could probably be built for a substantially lower cost than the Hubble Space Telescope. It would give us sharp images of exceedingly faint objects— ultra-remote galaxies, small planets, and so forth.

Our main information about distant stars and galaxies comes, of course, from the light we detect from them. But ordinary visible light is just a part of the electromagnetic spectrum. Beyond the blue end of the spectrum lies the ultraviolet, consisting of radiation with a shorter wavelength than blue light, carried by photons with more energy. Beyond the ultraviolet is the X-ray band, which corresponds to still more energetic photons. At the other end of the spectrum, red light merges into the infrared, the heat radiation that is emitted by objects not hot enough to glow visibly. At yet longer wavelengths are the microwave and radio bands.

If we think of an acoustic analogy, the ultraviolet and X-rays are the high notes; the infrared and radio waves lie at the other end of the keyboard. The whole electromagnetic spectrum emitted by cosmic objects ranges over more than a hundred octaves. In this analogy, visible light, from the red to the blue, is just a single octave. Optical observations alone therefore give us an incomplete perspective on the cosmic scene—narrow and insipid midrange harmonies rather than the broad range of frequencies that most cosmic objects actually radiate. This limitation became apparent as soon as the sky was scanned with crude radio antennas in the early days of radio astronomy. Some of the strongest emitters of cosmic radio noise could be readily identified: one, for instance, was the Crab Nebula. Others were extragalactic objects (involving, we now realize, energy generation around gigantic black holes) so remote that they could hardly be seen at all with optical telescopes. The most prominent features of the radio sky certainly weren't the same as the brightest visible objects. The physical processes that emit the radio waves, though now well understood, were not predicted.

Many types of radiation from cosmic objects get absorbed in air and don't penetrate down to a ground-based observatory. Space-based telescopes have opened up new techniques for observing the cosmos—for instance, by allowing astronomers to detect X-rays from space. The first X-ray detectors, mounted on rockets, each yielded only a few minutes of useful data before they crashed back to the ground. Like the Apollo program to land men on the Moon, X-ray astronomy derived its impetus from superpower rivalry in space technology and nuclear weapons. This technique spurted forward in 1970 when NASA launched the first satellite carrying X-ray detectors; it stayed in orbit, gathering data, for years rather than just a few minutes. This small satellite was built and operated by a research group led by Riccardo Giacconi, an Italian physicist who had settled in the United States. Through this project and its many successors, X-ray astronomy has proved itself to be a crucial new "window" on the universe. The improvement in cosmic X-ray telescopes is as dramatic as the advance from Galileo's "optic glass" to the 10-meter telescope, and it has all happened in less than thirty years.

Thermal X-rays are emitted by objects thousands of times hotter than the surfaces of ordinary stars; X-ray maps of the sky consequently highlight the hottest and most energetic objects in the cosmos. For instance, intense X-rays come from violently heated gas swirling into black holes. Huge holes of this type, each weighing as much as the total of millions or even billions of stars, lurk in the centers of galaxies. The radiation emitted from close to them carries clues to the most remarkable aspects of Einstein's theory—how space is distorted, and how time is stretched, by strong gravity.

Astrophysicists are especially interested in the most extreme and inclement cosmic conditions: violent explosions, jets of particles at 99.99 percent of the speed of light, flashes that disgorge in a few seconds far more light than the Sun emits in its 10-billion-year history. The cosmos offers us a laboratory where we can test the laws of nature under extreme conditions that cannot be achieved on Earth.

Looking Far Back

As described in chapter 3, laboratory knowledge of gravity, atoms, and their nuclei allows us to compute the life cycle of the Sun and stars. Now the gross properties of entire galaxies are also coming into focus. The stars in our Milky Way and Andromeda are orbiting in a disk, in a way that Newton would have readily grasped. But there is complicated "weather" in galaxies, too, churning up the interstellar gas and recycling it through successive generations of stars—this is how the atoms from the periodic table are built up from pristine hydrogen.

In their classic book, *Galactic Dynamics*, James Binney and Scott Tremaine make the nice point that galaxies are to astronomy what ecosystems are to biology. The atoms that we are made of come from all over our Milky Way Galaxy, but few come from other galaxies. But where did the original hydrogen come from? And why do galaxies exist? Why is our universe made up of these aggregates of gas and stars, each tens of thousands of light-years across? To answer these questions, we must probe far back to the first few minutes of cosmic history.

Just as geologists infer climatic history by drilling through successive layers of Antarctic ice, astronomers infer the history of our Galaxy by studying, and trying to date, the various populations of stars within it. But astronomers have an advantage over geologists: they can actually see the remote past by looking at galaxies so far away that their light was emitted billions of years ago.

Figure 4.3 depicts our world line and that of other galaxies. The only parts of space-time we know about are the

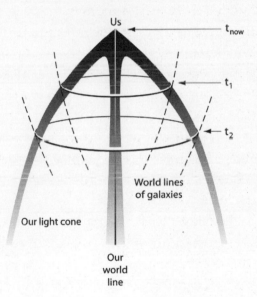

4.3

This space-time diagram shows our own world line and the world lines of some distant galaxies. We have direct evidence only about the shaded parts: the history of our own Galaxy (events near our own world line), and events along our past light cone. Because our universe is nearly homogeneous, conditions were similar everywhere on each time slice (depicted as horizontal planes), and we can make confident inferences about the intervening regions.

EXTRAGALACTIC PERSPECTIVE

shaded bits: our own world line and the parts along the past "light cone." In a completely random universe (resembling a mountain landscape rather than an ocean surface), distant regions might not resemble anything in our vicinity and might have quite different histories. The large-scale uniformity of our universe is crucial here. It is only because of the overall smoothness of the universe that we have grounds for believing that all of its parts have evolved similarly, and that a picture of galaxies, say, 5 billion light-years away is similar to a 5-billion-year-old snapshot of our own Galaxy and its neighbors.

The most detailed pictures of the sky that we have so far obtained came from week-long exposures with the Hubble Space Telescope. Each of these images shows a patch of sky so small that it would cover less than a hundredth the area of the full moon. Such a patch looks blank when viewed with an ordinary telescope. But it reveals many hundreds of faint smudges, each of them actually an entire galaxy, thousands of light-years across. They appear small and faint because they are so far away: their light began its travels 10 billion years ago. We see them as they were in the remote past, when all their stars were still young. Scientists at the Space Telescope Institute in Baltimore have now published several of these Hubble Deep Field images, and it turns out that patches in the northern and the southern sky look, statistically, just the same. This finding confirms the large-scale uniformity.

We have, of course, known since early in the twentieth century that our own solar system is several billion years old, so it should not surprise us that our expanding universe extends for billions of light-years. Still, it stretches our comprehension to contemplate these billions of galaxies, each capable of harboring millions of Earthlike planets. In that perspective, the amazing pictures from the Hubble Telescope, and from larger instruments on the ground, will impact as strongly on the public consciousness as the images of our Earth and the other planets that have been beamed back from space ever since the 1960s.

When we observe Andromeda, we may wonder if Andromedans are looking back at us, maybe with still bigger telescopes. Perhaps they are. But the remotest galaxies viewed in the Hubble Deep Field could not yet have evolved anything so advanced. We are viewing them at a very primitive stage, before there's been time for many stars to have completed their lives and for the stellar furnaces to have forged the atoms needed for complex chemistry. There is very little oxygen, carbon, silicon, and so forth to make even rocky planets, so there is scant chance that life has even started.

Friendly skeptics sometimes ask me: "Isn't it presumptuous for cosmologists to claim to explain anything about the vast cosmos?" My response is that what makes things hard to understand is how *complicated* they are, not how *big* they are. Under extreme conditions—inside the stars or in the hot early universe—everything breaks down to its simplest ingredients. A star is simpler than an insect. Biologists, tackling the intricate multilayered structures of butterflies and brains, face tougher challenges than astronomers.

I mention this disclaimer preemptively because in the next chapter I make assertions that might otherwise seem arrogant—for instance, that we can infer what happened in the first few seconds of cosmic history with 99 percent

EXTRAGALACTIC PERSPECTIVE

confidence, and that we can trace everything back to a Big Bang whose essential properties are quite simple to describe.

These advances bring new questions and mysteries into sharper focus. How did 13 billion years of evolution lead from such a simple recipe to our complex habitat where— here on Earth and perhaps on other worlds—atoms assemble into creatures able to ponder their origins? Maybe there are aliens in orbits around distant Suns who already know the answers. But for us, they are challenges for the new millennium.

5 Pregalactic History

Before Galaxies: Back to a Hot Beginning

The Hubble Deep Field images offer a portrait of the era when galaxies were newly forming. But what about still earlier cosmic history, before even the first star formed? Some seventy years ago, Georges Lemaître, a Belgian priest who was also a mathematician and an MIT graduate, pioneered the idea that everything began in a dense state. He called this the "primeval atom"; but that phrase never caught on. Nor did the word "ylem" introduced by the boisterous Russian-American, George Gamow. These coinages were usurped by

"Big Bang"—a flippant term that Fred Hoyle introduced in the 1950s as a derisive description of a theory he didn't like. Hoyle then favored the view that our universe was in a steady state: new atoms were being created all the time and agglomerated into new galaxies that filled in the gaps between the old ones so that, on average, everything always looked the same, despite the expansion. This theory—that our universe existed from everlasting to everlasting in a uniquely self-consistent state—was popular in England in the 1950s. It had the virtue of making very specific predictions; it was vulnerable to disproof, and it served as a goad to observers. But the voices of the articulate trio who invented it—Hermann Bondi, Thomas Gold, and Fred Hoyle—never really carried across the Atlantic. Nor were their views ideologically acceptable in the Soviet Union.

In the 1940s and 1950s, the most vocal advocate of the Big Bang was George Gamow. With his younger collaborators Ralph Alpher and Robert Hermann, he studied the physics of the hot dense beginning. Their calculations were corrected and refined by Sitoru Hayashi in Japan but attracted little interest at the time, mainly because theories about a dense beginning of the universe (if there was one) seemed inaccessible to observation, and hence unconstrained speculation.

Lemaître wrote: "The evolution of the universe can be likened to a display of fireworks that has just ended: some few wisps, ashes and smoke. Standing on a well-chilled cinder we see the fading of the suns, and try to recall the vanished brilliance of the origin of the worlds."[1] The evidence for this "vanished brilliance" dates from 1965, when an afterglow of our universe's hot, dense beginning was detected—certainly the most important cosmological advance of the last fifty years. This discovery, which gave the steady-state theory its coup de grâce, was made serendipitously by Arno Penzias and Robert Wilson, staff members at the Bell Telephone Laboratories in New Jersey, who were perplexed by excess "background hiss" in a sensitive antenna intended to pick up transmissions from artificial satellites. They announced this detection in a famous paper in the *Astrophysical Journal* with the undramatic title "Excess Antenna Temperature at 4080 Mc/s." Another paper in the same issue spelled out what it meant. This second paper was by the late Robert Dicke and his Princeton colleagues James Peebles, Paul Roll, and David Wilkinson, who had been planning to detect the radiation themselves. They immediately realized that Penzias and Wilson had scooped them in a momentous discovery, and they quickly made follow-up measurements of their own.

Intergalactic space is not completely cold but is slightly warmed by all-pervasive microwaves with no apparent source; this radiation suffuses our entire universe. Its intensity at different wavelengths, when plotted on a graph, traces out what physicists call a "black body" or "thermal" curve. This particular curve is expected if the radiation has come into balance with its environment, as happens deep inside a star or in a steadily burning furnace. Evidence for this special spectrum mounted up during the twenty-five years after the radiation was discovered, but in 1990 John Mather and his colleagues used the Cosmic Background Explorer (COBE) satellite to make truly remarkable measurements with errors smaller than the thickness of the line on the graph (fig. 5.1). This confirms beyond any reasonable doubt that everything—all the

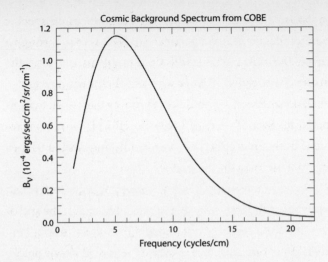

Cosmic Background Spectrum from COBE

y-axis: B_V (10^{-4} ergs/sec/cm^2/sr/cm^{-1})

x-axis: Frequency (cycles/cm)

5.1

The background radiation, as measured by the COBE satellite. The shape matches a black-body spectrum to a precision of one part in 10,000.

stuff that galaxies are now made of—was once a compressed gas, hotter than the Sun's core. The expansion has cooled and diluted the radiation and stretched its wavelength. But this primordial heat—the afterglow of creation—is still around: it fills all of space and has nowhere else to go! Most of us see it every day, as it causes one percent of the interference on our television screens.

Penzias and Wilson were radio astronomers, with expertise in electronics rather than cosmology. Their success stemmed from persistence and technical skills. It wasn't surprising that it took others to convince them of what their discovery meant. In fact, Wilson has recalled that the full import of his achievement really sank in only when he read a report of it in the *New York Times*. Many of us, at a lower level, have similar experiences. Any researcher has to focus on specific technical details, but the occupational risk is that, through a narrow focus on tractable bite-sized problems, one loses the broader perspective. That is why—in their own interests—professional scientists should try to convey their work to non-specialists. Even if we do it badly, the effort is salutary: it reminds us that our efforts are worthwhile only insofar as they help to illuminate the big picture.

For its first few minutes, our universe was far hotter than the center of the Sun—hotter even than big stars at the end of their lives, and certainly hot enough for nuclear fusion. Fortunately for our existence, however, it cooled down before there had been enough time to "process" everything into iron, the most tightly bound nucleus. If our universe had stayed hot for longer (or if the reactions had happened faster), there would have been no nuclear fuel left to power the stars. But the only thing that happened is that about 23 percent of the hydrogen turned into helium, leaving a bit of heavy hydrogen (deuterium) as an intermediate product (fig. 5.2). These proportions agree with what astronomers measure. Apart from a trace of lithium, no elements higher up on the periodic table—no carbon, oxygen, etc.—emerged from the immediate aftermath of the Big Bang.

The expanding universe, initially blazingly hot and bright, took hundreds of thousands of years to cool down to the temperature of the Sun's surface. At that stage, electrons and ions combined into neutral atoms that no longer scattered the radiation. The "fog" lifted and the universe became transparent. The primeval light then shifted into the infrared, and the universe became literally dark until the first stars

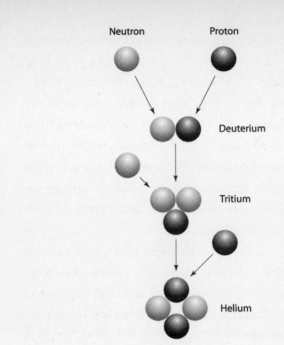

5.2
Nuclear reaction networks for synthesis of deuterium and helium in the first three minutes after the Big Bang.

formed and lit it up again. The microwaves have been traveling uninterruptedly since the universe was less than a thousandth of its present age; they are direct "messengers" from an era before any stars or galaxies had formed.[2]

Dark Matter

We don't know whether or not our universe is infinite, but it is manifestly very large. Enough atoms of hydrogen and helium emerged from the fireball phase to make all the stars in all the galaxies; there are at least 10^{78} atoms within range of our telescopes. And it contained quanta of radiation, or photons, in far greater numbers still—about 2 billion photons for each atom, which is why we often refer to the *hot* Big Bang.

Something else survives from the hot early stages: the mysterious dark matter. Astronomers have discovered that galaxies, and even entire clusters of galaxies, would fly apart unless they were held together by the gravitational pull of between five and ten times more material than we actually see. There are many types of observations that support this conclusion. I'll mention just two.

The first evidence comes from disk galaxies, like our own Galaxy or Andromeda. Stars and gas circle around the central hub of such galaxies, at a speed such that centrifugal forces balance the gravitational pull toward the center. This is analogous—though on a hugely larger scale—to the way the Sun's gravitational pull holds the planets in their orbits. Radio astronomers can detect clouds of cold hydrogen, orbiting far out beyond the limit of the visible disk of stars; hydrogen atoms emit at a characteristic radio wavelength of 21 centimeters, and the speed of the clouds can be inferred from the Doppler shift of this emission. If these outlying clouds were feeling the gravity of the stars and gas in the galaxy, then the farther out they were, the slower they would be moving—for the same reason that Pluto is moving more slowly around the Sun than Earth is. But that's not what is happening, since clouds at different distances from the galaxy all orbit at more or less the same speed. If, in our own solar system, Pluto were moving as fast as Earth, we would have to conclude that there is a shell of material outside Earth's orbit but still within Pluto's. Likewise, the high speed of these outlying clouds tells us that there is more to galaxies than meets the eye. The entire luminous galaxy—an assemblage of stars and glowing gas—must be embedded in a dark halo, several times heavier and more extensive.

The speeds of stars and gas in galaxies, incidentally, are no more than a thousandth of the speed of light. This means that the corrections introduced into Newton's theory by Einstein's general relativity are very minor, as they also are for planetary orbits.

And there's pervasive dark matter on still larger scales. The Swiss-American astronomer Fritz Zwicky argued in the 1930s that the galaxies in a cluster would disperse unless they were restrained by the gravitational pull of dark matter. He proposed that gravitational lensing—the bending by gravity of light rays from objects behind it—could reveal dark matter's presence. This technique has now, sixty-five years later, borne fruit, and it is sad that Zwicky didn't live long enough to see images like the one shown on the cover of this book, which depicts a big cluster of galaxies about a billion light years away. The image also shows numerous faint streaks and arcs: each is a remote galaxy, several times farther away than the cluster itself, whose image is, as it were, viewed through a distorting lens. Just as a pattern on background wallpaper looks distorted when viewed through a curved sheet of glass, the gravity of the cluster of galaxies deflects the light rays passing through it. The visible galaxies in the cluster contain only 10 to 20 percent as much material as is needed to produce these distorted images—evidence that clusters, as well as individual galaxies, contain five to ten times more mass than we actually see.

The main case for dark matter rests on applying Newton's law of gravity on scales millions or even billions of times larger than our solar system, which is of course the only place where it has been reliably tested. It is proper to be cautious about such a huge extrapolation; some have even suspected that we are indeed being misled, and that gravity grips more strongly at large distances than standard theories predict. In this context the gravitational lensing is important because it corroborates the other evidence, yet it is based on rather different physics—Einstein's rather than Newton's.

There's really no reason to demur at the idea that most of the stuff in the universe may be dark: why should everything in the sky shine any more than everything on Earth does? What we actually see could be a small and atypical fraction of what is out there. But what could the dark matter be? It is embarrassing to cosmologists that 90 percent of the universe is unaccounted for. Earlier candidates for the missing matter included very faint stars and the dead remnants of stars. But surveys are now much more sensitive, and if there were enough objects of these types to account for all the dark matter, it is unlikely that they would have escaped detection.

We now strongly suspect that the dark matter cannot consist of anything that is made from ordinary atoms. The favored view is that it consists of swarms of particles that have so far escaped detection because they have no electric charge, and because they pass straight through ordinary material with barely any interaction.[3] There is still boisterous debate about exactly what these mysterious and elusive entities could be; there are no firm candidates. On the other hand, physicists have theorized about many types of particles that could have been created in the ultra-hot initial instants after the Big Bang and survive until the present.

Thousands of these particles could be hitting us every second, but they almost all pass straight through us, and through any laboratory. Sometimes one of them collides with an atom, however, and sensitive experiments might detect the consequent recoil when this happens within (for instance) a lump of silicon. Several groups around the world have taken up this challenge. It requires delicate equipment, cooled down to temperatures near absolute zero and deployed deep underground to reduce the background signal from cosmic rays and so forth.

One of the experimental difficulties is that other kinds of particles (for instance, decay products from radioactivity in rock) can cause similar signals. But genuine dark matter would have a distinctive signature, which would tell us that it came from our Galaxy and not just from Earth. The Sun moves steadily through the swarm of particles that make up our galactic halo, but Earth moves around the Sun, and so our speed through the halo varies in a predictable way during each year: highest in June and lowest in December. In 1999 an Italian group, using equipment deep under the Gran Sasso mountain in the Apennines, actually claimed to have discerned just such a seasonal modulation in their data. But it wasn't confirmed. Dedicated scientists will have to improve their techniques further and stay down in their tunnels and mineshafts for a few years more before they can detect an unambiguous signal. But even then success isn't assured as there are still many uncertainties about the particles that make up the dark matter: they might be individually too light to produce a detectable recoil, or they might be too heavy and rare.

The intellectual stakes are high. Dark matter is the No. 1 problem in astronomy today, and it ranks high as a physics problem, too. If we could solve it—and I'm optimistic that we will within the next decade—we would know what our universe is mostly made of, and we would discover, as a bonus, something quite new about the microworld of particles. Moreover, as discussed in the next chapter, the amount of dark matter affects the cosmic long-range forecast—how the universe will be expanding tens of billions of years hence.

We're reconciled to the post-Copernican idea that we don't occupy a central place in our universe. But now our cosmic status must seemingly be demoted still further. Particle chauvinism has to go: we're not made of the dominant stuff in the universe. We, the stars, and the visible galaxies are just traces of sediment—almost a seeming afterthought—in the cosmos; something quite different (and still unknown) controls its large-scale structure and eventual fate.

From Simplicity to Complexity: The Role of Gravity

Our cosmic habitat displays great contrast and variety. There is a huge range of temperatures. The stars have blazing surfaces (and still hotter centers), but the dark sky is warmed just to 2.7 degrees above the absolute zero of temperature by the microwave afterglow from the Big Bang. And there is a huge range of densities too: some atoms are still spread through intergalactic space, with less than one in each cubic meter;

others are assembled into galaxies, stars, and planets—and at least one biosphere.

All this intricate complexity emerged from an amorphous fireball that takes no more numbers to describe it than a single atom does. This might seem to violate a hallowed physical principle—the second law of thermodynamics—which describes an inexorable tendency for patterns and structures to decay or disperse: things tend to cool if they are hot, and to warm up if they are cold; ordered states get messed up, but not the reverse. Entropy, a measure of this disorder, never decreases.

What is the answer to this seeming paradox? The force of gravity is crucial—in particular, the way this force amplifies small initial density contrasts in an expanding universe. Any patch that starts off slightly denser than average, or is expanding slightly slower than average, would decelerate more because it feels extra gravity. Its expansion lags farther and farther behind, until it eventually stops expanding and separates out as a gravitationally bound system. This process is, we believe, what allowed galaxies of stars to form about a billion years after the Big Bang.

Astrophysicists can study this process by simulating virtual universes in their computers, showing how slight initial overdensities evolved, under the action of gravity, into galaxies, which in turn grouped themselves into clusters.

Figure 5.3 shows six frames from a simple movie showing a cube large enough to make dozens of galaxies. The expansion is scaled out, so all boxes look the same size. You can clearly see the incipient structures unfolding and evolving. These pictures show only the dark matter, which has the

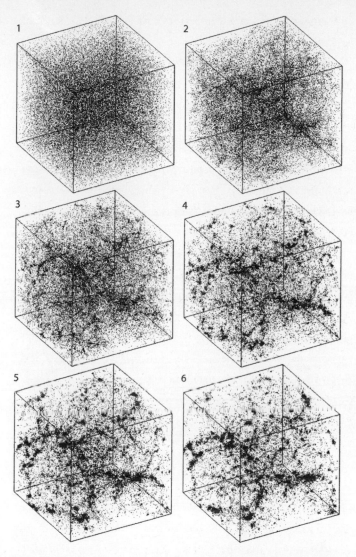

5.3
Six frames from N-body simulations, showing how the density contrasts grow and structure develops. In these pictures, the earlier and later boxes have been rescaled so that, despite the expansion, all look the same size.

dominant gravitational effect. Smaller structures merge together, producing protogalaxies. Gas would be pulled into these galaxy-scale clumps of dark matter, where it would cool and condense into "droplets" that would turn into stars. The new galaxies would then aggregate into clusters. Movies of this kind depict the emergence of galaxies at about 16 powers of 10 faster than it actually happened. The starting point of these calculations is an expanding universe containing atoms, radiation, and dark matter.

Newton himself pondered the origin of cosmic structure. He did not have our modern conception of the actual scale of our universe, nor of course did he know that it was expanding; but in a letter to Richard Bentley, a classical scholar who was master of Trinity College at Cambridge, he imagined how material in an infinite static universe might condense into stars:

> If the matter were evenly dispersed throughout an infinite space . . . some of it would convene into one mass and some into another, so as to make an infinite number of great masses, scattered at great distance from one another throughout all that infinite space. And thus might the sun and fixed stars be formed.

He went on to add the proviso "supposing the matter to be of a lucent nature." But it would be wishful thinking to see this as a premonition of dark matter.

The computer simulations successfully mimic the gross properties of actual galaxies—the characteristic sizes and shapes, the proportions that are disklike and the proportions that are elliptical—and the way they are clustered. And they offer, incidentally, a further reason for believing in dark matter: the outcome of these simulations does not match our actual universe so well if everything is assumed to be made of atoms and there is no dark matter to provide extra gravity.

The finer details, the "weather patterns" within each galaxy, are much harder to model. The trend away from uniformity toward ever larger contrasts in temperature and density continues after stars form, because gravity induces a seemingly perverse relation between heat, energy, and temperature. Suppose the fuel supply in the Sun's center were switched off. It would gradually deflate as energy leaked away. But the center would end up *hotter* than before: gravity pulls more strongly at shorter distances, and the core has to heat up in order to provide enough pressure to balance the greater force pressing down on it. Ordinary substances (water, for instance) have a characteristic called "specific heat" that measures how much energy they soak up when their temperature is raised. Objects held together by gravity have a negative specific heat—they get ever hotter, ever denser, as they lose energy.

Once stars and planets have formed, subsequent events on some of those planets could be more complex than anything that came before. Basic physics tells us that no "heat engine" can operate, and no complexity can emerge, if everything is in thermal equilibrium: there must be some regions that are hotter than others. Planetary biospheres are energized by light from the central star. This "high grade" energy drives photosynthesis, and the waste heat is radiated into cold interstellar space. Biological evolution is sensitive to accidents—climatic changes, asteroid impacts, epidemics, and so forth—so that, if Earth's history were to be rerun, its biosphere would end up quite different. The same would be true on any other planet.

Incidentally, astronomers habitually refer to the "evolution" of stars and galaxies, but a closer parallel with biological usage would be achieved if the word "development" were used instead. The entities studied by astronomers—stars, galaxies, and the like—manifest a trend toward enhanced complexity and differentiation, just as a growing animal or plant does. But stellar and galactic astronomy offer no analogue of Darwinian evolution by natural selection.

Cosmic Texture

The universe cannot have started off perfectly smooth and uniform. If it had, it would now contain hydrogen and helium gas so rarefied that there would have been less than one atom in each cubic meter everywhere. It would have been cold and dull: no galaxies, therefore no stars, no periodic table, no complexity, certainly no people.

But because of the "contrast enhancement" that's induced by gravity during the expansion, even a slight initial nonuniformity could change all that. The amplitude of these nonuniformities can be described by a simple number Q— the energy difference between peaks and troughs in the density, expressed as a fraction of the total energy (Einstein's mc^2) of the material. Q determines the scale of the biggest structures in the universe, with larger values of Q leading to a "lumpier" universe. The computer models suggest that Q has to be about 0.00001 in order to account for present-day galaxies and clusters. The smallness of Q implies that, in terms of gravity, our universe is as smooth as the Earth would be if the highest mountains or waves were only 50 meters high.

These irregularities, or ripples, reveal themselves in another way: they make the background slightly warmer in some parts of the sky and slightly cooler in others. The COBE satellite mapped the temperature of the whole sky and found temperature variations of one part in 100,000. This measurement was a real technical triumph. The radiation, less than 3 degrees above the absolute zero of temperature, is a hundred times cooler than Earth and its atmosphere, and the temperature difference between hot and cold patches on the sky are 100,000 times smaller still.

The COBE measurement also corroborated the idea that cosmic structures emerged via gravitational instability. It showed that the hot early universe was indeed pervaded by ripples with just the amplitude that was needed, according to the computer calculations, to account for our universe's present structure.

The fluctuations must have been imprinted very early, along with the mix of ingredients in our universe. Whatever caused the Big Bang seems to have left it ringing or vibrating, but there is still no understanding of what fixes the amplitude of these vibrations—in other words, the value of Q. It is interesting, however, that neither a much smoother nor a much rougher universe would offer such a benign habitat.

If Q were much smaller than 0.00001 (or 10^{-5}), galactic "ecosystems" would never form: aggregations would take longer to develop, and their gravity would be too weak to retain gas. A very smooth universe would remain forever dark and featureless, even if its initial mix of atoms, dark matter, and radiation were the same as in our own universe. On the other hand, a rougher universe, with Q much larger than 0.00001, would be turbulent and violent. Lumps far bigger than galaxies would condense early in its history. They would not fragment into stars: instead, they would collapse into vast black holes, each much heavier than an entire cluster of galaxies in our universe. Even if galaxies managed to form, the stars in them would be packed so close together that any planetary systems would be buffeted by passing stars.*

How Credible Is the Big Bang Theory?

The evidence that our universe expanded from a hot dense state has firmed up over the years. The Big Bang theory has lived dangerously for decades—and survived. Several things could have been discovered during the 1980s or 1990s that would have discredited the hypothesis, but they weren't. For example, any object whose helium content was zero, or at any rate well below 23 percent, would have been fatal because extra helium made in stars can readily boost helium above its pregalactic abundance, but there seems no way of converting all of the helium back to hydrogen. Or the background radiation measured so accurately by the COBE satellite (shown in fig. 5.1) might have turned out to have a spectrum that differed from the expected black body or thermal form. Excess

---

*For similar reasons, stable planetary systems would be less likely to exist near the center of our own Galaxy, where the stars are in a dense-packed swarm, rather then being as widely dispersed from each other as they are in the Sun's immediate locality.

radiation at the shortest (millimeter) wavelengths could have been due to diffuse dust or other sources. But if there were, at some wavelengths, a deficit compared to the standard blackbody radiation, that would have been a major mystery.[4]

The Big Bang theory deserves to be taken at least as seriously as anything geologists or palaeontologists tell us about the early history of our Earth: the inferences that Earth scientists make are just as indirect (and less quantitative). The theory's survival gives me (and I suspect most cosmologists today) 99 percent confidence in extrapolating right back to the first few seconds of cosmic history.

I would prudently leave the other one percent for the possibility that our satisfaction is as illusory as that of a Ptolemaic astronomer who had successfully fitted some more epicycles. Cosmologists are sometimes chided for being often in error but never in doubt.

My old guru, Fred Hoyle, inventor of the "steady state" theory, still is not reconciled to all this: he has come part of the way and believes in a "steady bang"—a cosmos undergoing a very slow expansion, on which a more rapid oscillation is superimposed. The physicist Max Planck claimed that theories are never *completely* abandoned until their proponents die. And Fred is fortunately, at the time I'm writing this, still very much alive.

(Ironically, the steady statesmen may turn out to have been not 100 percent wrong, but merely too limited in envisioning the overall scale of the cosmos. I will mention in the final chapters the concept that our Big Bang could be one of many "bangs" popping off in a cosmos that on some gigantic scale persists in an eternal self-reproducing stationary state.)

Figure 5.4 is a time chart showing the Big Bang, right back to the beginning. I should emphasize very strongly that my 99 percent confidence extends back to one second—perhaps even to a millisecond—but certainly not much farther back than that. This limit arises because a backward extrapolation into the first tiny fraction of a second leads to densities and temperatures far higher than can be reproduced or studied on Earth, so we have no firm basis for any inferences

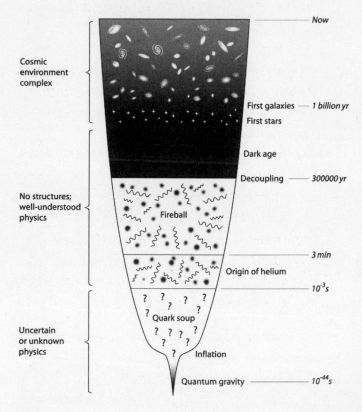

5.4
Time chart of the universe.

about the conditions that then prevailed. After the first millisecond, the particles would have been moving and colliding with energies that can be readily reproduced in the lab. The background radiation and the helium and deuterium are fossils of that era, and the concordance between measurements and theory corroborate the extrapolation that far back.

In brief, we can distinguish three eras of cosmic history.

1. *The first millisecond.* Here, there are uncertainties in the basic physics, which get more serious the farther back we go, since we gradually lose our foothold in experiment.
2. *From a millisecond to some millions of years.* This part is the easy bit. The physics is well known, and everything is still smoothly expanding. But the simplicity ends when the first structures condense out.
3. *The "recent" universe.* This era is intractable, not because we don't understand the physics, but for the same reason that other environmental sciences such as meteorology are difficult. But the whole evolving fabric depends on how the conditions were set up. The key numbers like Q and the mix of atoms, radiation, and dark matter are all legacies of the exotic and uncertain physics of the first era.

We are witnessing a crescendo of discoveries that we can be confident will continue throughout the next decade because of a confluence of developments.

1. *Fluctuations in the background temperature.* These fluctuations are caused by the "embryos" of galaxies and

clusters. Within the next few years, two satellites, NASA's MAP satellite and the European Space Agency's "Planck-Surveyor," will have covered the whole sky with far higher resolution than COBE achieved.

2. *The high-redshift universe.* The Hubble Space Telescope has fulfilled its potential; the two Keck Telescopes are on-line (now joined by the VLT in Chile and several other instruments with mirrors more than 8 meters in diameter). New X-ray telescopes in space, and radio arrays on the ground, now offer greater sensitivity. A decade from now, still larger telescopes, both in space and on the ground, will carry the enterprise still farther than current instruments can achieve.

3. *Large-scale clustering.* Big surveys will catalog millions of galaxies, permitting a whole battery of sensitive statistical tests, which should pin down more of the details of how galaxies formed.

4. *Dramatic advances in computer technology.* These enable us to stimulate, with ever-improving realism, how galaxies and stars form and develop. We can even do "virtual experiments" to discover what happens when stars explode or when galaxies collide.

5. *New concepts in fundamental physics.* As I will discuss in chapter 9, new concepts offer the hope (unless current euphoria is completely dashed) of putting the ultra-early universe, the first tiny fraction of a second, on as firm a footing as the later eras.

6 Black Holes and Time Machines

Complete Gravitational Collapse

Ever since the beginning, gravity has been making our universe less and less uniform and building up ever-larger contrasts of density and temperature. In the end, gravity overwhelms all the other forces in stars, and in anything larger, even though the effects of rotation and nuclear energy delay its final victory.

There are some entities in which gravity has already triumphed over all other forces. These are black holes—objects that have collapsed so far that no light or any other signal can escape them, but that nonetheless leave imprints, distortions of space and time, frozen in the space they've left.

An astronaut who ventured too close to a black hole would pass into a region from which there is no return and from where no light signals can be transmitted to the external world; it is as though space itself were being sucked inward faster than light moves through it. An external observer would never witness the falling astronaut's final fate: any clock would appear to run slower and slower as it fell inward, into the hole, so the astronaut would appear impaled at a horizon, frozen in time.

The Russian theorists Yakov Zeldovich and Igor Novikov, who studied how time was distorted near collapsed objects, coined the term "frozen star" for such objects. Zeldovich, one of the last polymaths of physics, holds a prominent place in modern cosmology. He was a dynamic and charismatic personality; from the 1960s onward, his research school in Moscow spearheaded many key discoveries even though cosmology and relativity had previously been ideologically tainted in the USSR).[1] The term "black hole" itself was not coined until 1968, when John Wheeler described how an infalling object "becomes dimmer millisecond by millisecond . . . light and particles incident from outside . . . go down the black hole only to add to its mass and increase its gravitational attraction."[2]

Black holes, the most remarkable consequences of Einstein's theory, are not just theoretical constructs. There are huge numbers of them in our Galaxy and in every other galaxy, each being the remnant of a star and weighing several times as much as the Sun. There are much larger ones, too, in the centers of galaxies. Near our own galactic center, stars are orbiting ten times faster than their normal speeds within a galaxy. They are feeling, close up, the gravity of a dark object, presumably a black hole, as heavy as 2.6 million suns. Yet our Galaxy is poorly endowed compared to some others, in whose centers lurk holes more massive than a billion suns, betraying their presence by the high-speed motions of surrounding stars and gas, induced by their gravitational pull.

Black holes are among the most exotic entities in the cosmos. But they are actually among the best understood. They are constructed from the fabric of space itself and are as simple in structure as elementary particles. A newly formed black hole quickly settles down to a standardized stationary state characterized by just two numbers: those that measure its mass and its spin. (In principle, electric charge is a third such number, but stars can never acquire enough electric charge for this factor to be relevant to real collapses.) The distorted space and time around black holes is described exactly by a solution of Einstein's general relativity equations that was first discovered in 1963 by Roy Kerr, a mathematician who later forsook research to become an internationally recognized bridge player. In general, macroscopic objects seem more and more complicated as we view them closer up, and we can't expect to explain their every detail; but black holes are an exception to this rule.

Viewed from outside, no traces remain that distinguish how a particular hole formed, nor what kind of objects it swallowed. The great Indian astrophysicist Subrahmanyan Chandrasekhar was deeply impressed by this realization, aesthetically as well as scientifically: "In my entire scientific life," he wrote, "the most shattering experience has been the realization that an exact solution of Einstein's equations of general relativity, discovered by the New Zealand mathematician Roy Kerr, provides the absolutely exact representation of untold numbers of massive black holes that populate the Universe."[3] Roger Penrose, the theorist who perhaps did most to stimulate the renaissance in relativity theory that occurred in the 1960s, has remarked: "It is ironic that the astrophysical object which is strangest and least familiar, the black hole, should be the one for which our theoretical picture is most complete."[4] The discovery of black holes thus opened the way to testing the most remarkable consequences of Einstein's theory.

I mentioned in chapter 1 NASA's ambitious Terrestrial Planet Finder mission—an array of mirrors in space that will be able to detect Earthlike planets orbiting other stars. When that discovery has been achieved, NASA's next challenge could be to image the gas swirling down into a black hole. It would look like the simulation in figure 6.1. We should not hold our breath waiting for a real picture like this, but we can infer a lot about the hole before we have an actual image. The radiation from such objects comes primarily from hot gas swirling downward into the gravitational pit. Instruments already in space can collect all the emitted radiation and measure its spectrum. Any spectral feature would display huge Doppler effects, because the gas is moving at a large fraction of the speed of light. There is also an extra redshift—the so-called gravitational redshift—resulting from each photon loss of energy as it climbs out from a region where gravity is so strong. Measurements of this radiation (mainly X-rays rather

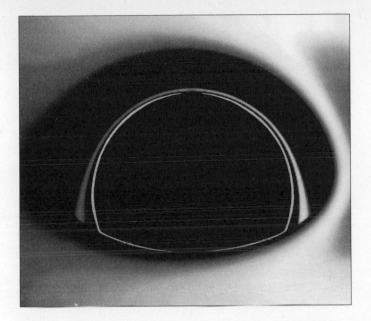

6.1
Gas swirling into a black hole. This sketch depicts what would be
seen when one takes account of Doppler shifts and light-bending
due to strong gravity.

than visible light, because the gas is so hot) can probe the flow
very close to the hole and diagnose whether the "shape of
space" around it agrees with what theory predicts.

These black holes interest astronomers because the flow
patterns and magnetic field around them generate some of
the most spectacular pyrotechnics in the universe. But they
challenge basic physics as well. Around any black hole is a
horizon, a surface shrouding from view an interior from
which not even light can escape. A hole's size is proportional
to its mass: if the Sun became a black hole, its radius would be
3 kilometers, but some of the supermassive holes in galactic

centers are as big as our whole solar system. If you fell inside one of these monster holes, you would be treated to several hours of leisurely observation before you approached the very center, where increasingly violent tidal forces would shred you apart. Right at the center, you, or your remains, would encounter the singularity where the physics transcends what we yet understand. The new physics that we'll need is the same that governs the initial instants of the Big Bang.

Fast-Forward (and Backward?) in Time

Good science fiction should respect the fundamental constraints of physical laws. In that spirit, it is worth mentioning that an observer could, in principle, observe the far future in what, subjectively, seemed quite a short time. According to Einstein, the speed of a clock depends on where you are and how you're moving. If your subjective clock ran very slowly compared to the cosmic clock, you could travel "fast forward" into the future. This would happen if you were moving at a velocity close to the speed of light. Furthermore, strong gravity would distort time: clocks on a neutron star would run 20 or 30 percent slower. Near a black hole, the distortions would be even greater. If you were to fall into a hole, your future would be finite: you would be ripped apart—spaghettified—by ever more violent gravitational forces. However, a more prudent astronaut who managed to get into the closest possible orbit around a rapidly spinning hole without falling into it would also have interesting experiences: space-time is so distorted there that his clock would run arbitrarily slow and he could, therefore, in a subjectively short period, view an immensely long future timespan in the external universe.

This elasticity in the rate of passage of time may seem counter to our intuition. But such intuition is acquired from our everyday environment (and perhaps, even more, that of our remote ancestors), which has offered us no experience of such effects. Few of us have traveled faster than a millionth of the speed of light (the speed of a jet airliner); we live on a planet where the pull of gravity is 1000 billion times weaker than on a neutron star. But time dilation entails no inconsistency or paradox.

More problematic, of course, would be time travel back into the past. More than fifty years ago, the great logician Kurt Gödel discovered that the theory of general relativity did not in itself preclude a time machine. He discovered a valid solution of Einstein's equations that described a bizarre universe where some of the worldlines were closed loops—in other words, you could come back into your own past. But Gödel's solution was not realistic: it described a universe that was rotating and not expanding.

Other theoretical examples of systems that seem to obey the laws of physics but which allow closed loops in time have been proposed. For example, Princeton theorist Richard Gott showed that a time machine could be constructed from two so-called cosmic strings—long microscopically thin tubes of hyperdense material, heavy enough to distort space. Gott and his colleague Li-Xin Li also devised a cosmological model even stranger than Gödel's in which an entire universe, with a finite life cycle, traces out a loop in time so that its end is also its beginning.

One much-discussed design for a time machine involves a "wormhole": two black holes linked together by a tunnel or "spacewarp." The tunnel could exist only if it were made of a substance that has a very large negative pressure (or tension). Theorists speculate that exotic stuff of this kind did exist in the early universe, but even if such material still existed, the mass needed in order to make a wormhole wide enough to be comfortably traversed by a human would be 10,000 times that of the Sun!

Gödel's discovery and its aftermath opened up a debate. Is there a further law of physics, more restrictive than Einstein's equations, that rules out such effects? One might call it a "chronology protection law." Or could a time machine in principle exist? Such an artifact plainly still lies in the hypothetical reaches of science fiction, but we can still ask whether the barriers to constructing a time machine are "merely" technological, or whether there is a fundamental physical law that prohibits them. (To clarify the distinction, most physicists would say that a large spaceship traveling at 99.99 percent of the speed of light is in the first category, but one that travels faster than light is in the second.)

The events on the time loop must close up self-consistently, as in a movie whose last scene recapitulates its first. Paradoxes arise if you come back into the past and undo something that was a precondition of your existence: for instance, murdering your grandmother in her cradle would raise issues of logical consistency, not just of ethics. Time travel makes sense only if some law of nature precludes inconsistency of this kind. The implication that there must be "time police" to constrain our free will might seem paradoxical. But I am convinced by the robust retort of Igor Novikov, a leading physicist who has explored these ideas, that physical laws already constrain our choices: we cannot, for instance, exercise our free will by walking on the ceiling. The prohibition on violating the consistency of a time loop is, in a sense, analogous.

Even if a time machine could be built, it would not enable us to travel back prior to the date of its construction. So the fact that we have not been invaded by tourists from the future may tell us only that no time machine has yet been made, not that it is impossible.

The next chapters describe what we can infer about the very beginning of our universe, and about its far future, without leaving the comforts of our home planet.

PART II The Beginning and the End

7 Deceleration or Acceleration?

Limits to Prediction

In August 1999, a total solar eclipse was visible from southwest England. I viewed it from Cornwall through intermittent clouds. For me it was simply an environmental experience, shared with thousands of New Age cultists, astrology devotees, and the like. But the spectacle triggered some simpleminded thoughts.

It reminded me, first, that astronomy is by far the oldest quantitative science. Eclipses could be predicted, at least approximately, in the first millennium B.C. For several centuries, the Babylonians recorded celestial events on cuneiform tablets, and thousands of these records can now be seen in the British Museum. They stretched over a long enough timespan to reveal subtle patterns—particularly an eighteen-year repetitive cycle—which could be extrapolated forward to predict when future eclipses were likely to occur. Such predictions were feasible for lunar eclipses, which are observable from half the Earth's surface, in contrast to solar eclipses, where "totality" occurs only along a narrow strip. Such predictions required no insight into how the Sun and Moon actually moved—only a faith in the regularity of nature.

It was not until the seventeenth century that substantial advances were made. By that time, astronomers such as Edmund Halley understood the layout of the solar system, and the eighteen-year cycle was realized to be due to a wobble in the plane of the Moon's orbit. Halley is famous for his insight that the comet he saw in 1682 was the same one that others had also seen in 1531 and 1607. He did not live to see its predicted return on schedule in 1758, though he had the good luck to see two total eclipses of the Sun in England during his lifetime, and he had predicted them both. His predictions of the "totality strip" were better than the ancients could have made. But more important was a qualitative advance: Halley, unlike the Babylonians, based his predictions on the kind of insight that we could properly call a scientific explanation.

Such an explanation, of course, removes any mystery and irrational dread. For example, a few weeks after Europe experienced the August 1999 eclipse, major earthquakes occurred in Turkey and Greece; in earlier centuries it would have been natural to treat these as causally linked, whereas we now understand eclipses and earthquakes well enough to realize that a causal link is unlikely.

Eclipses are atypical among natural phenomena in being highly predictable: we can now forecast them to the nearest second. But earthquakes can't be reliably predicted. Nor, of course, can the weather: when I went to Cornwall I didn't know whether it would be clear or cloudy. This unpredictability does not reflect the greater competence of astronomers over seismologists and meteorologists: it illustrates the crucial distinction between prediction and understanding. Eclipse forecasts, even centuries in advance, are almost as accurate as the input data. But weather forecasts are not because atmospheric flow patterns are unstable: the uncertainty increases because even a tiny tweak can make a big difference to the outcome, as is the case when one balances a pencil on its end.

But people still crave explanations even when there is no underlying understanding about what's going on. A chill is blamed on drafts, damp clothing, or some other conjectured cause. Likewise, erratic stock market movements always find a ready explanation in the next day's financial columns: a price rise is attributed to sentiment that "pessimism about interest rate increases was exaggerated," or to the view that "company X had been oversold." Of course these explanations are always a posteriori: commentators could offer an equally ready explanation if a stock had moved the other way.

Even when they're understood, most phenomena are still unpredictable, like the weather. The clockwork of the heavens is actually the exception rather than the rule. Newton was lucky to have hit on one of the very few things in nature that *are* predictable a long way into the future. Clean predictions are limited to the very small (the behavior of single particles, or simple chemicals) and the very large (celestial orbits and so forth). The most intractable complexity manifests itself on the intermediate scales of everyday experience—in structures large enough to contain many layers of structure, but not so large as to be crushed by gravity.

The subatomic world is simple—it seems to be describable by a few equations. On the largest scales also, our universe is simple: its gross properties are defined by just a few numbers: its expansion speed, the mean density, and so forth. An overall uniformity prevails, and gravity holds sway.

The Long-Range Cosmic Forecast?

If indeed the grand astronomical phenomena are in some respects simpler than everyday events, can we make some cosmic predictions? Can we even offer a long-range forecast for our expanding universe?

In about 5 billion years, the Sun will die, and the Earth with it. At about the same time, give or take a billion years, Andromeda may crash into our Galaxy. But will the universe go on expanding forever? Or will our firmament eventually recollapse into a Big Crunch? Space is already being punctured by the formation of black holes. Are these perhaps the precursors of a crunch that will engulf us all? (See fig. 7.1.)

The answer depends on how much the cosmic expansion is being decelerated. Everything exerts a gravitational pull on everything else, and it is easy to calculate that it would take the mass of only five hydrogen atoms per cubic meter to

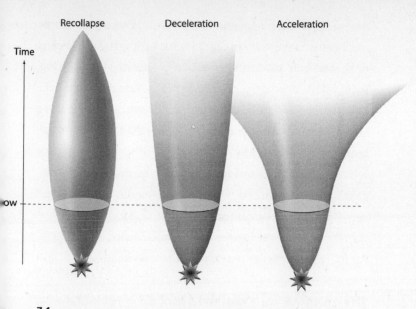

Recollapse   Deceleration   Acceleration

Time

ow

7.1
Trajectories for the far future of our universe.

bring the expansion to a halt—unless some other force inter-
venes. This doesn't sound like much. But if all the galaxies
were dismantled and their constituent stars spread uniformly
through space along with all the gas, they would make an
even emptier vacuum—one atom in a volume of 10 cubic
meters. There seems to be a similar amount of material in diffuse intergalactic gas, but even when that is added into the
mix, the resulting density amounts to only 0.2 hydrogen
atoms per cubic meter. This is equivalent to a few grains of
sand in the volume of Earth; or to a small asteroid, a few hundred meters across, in a box big enough to contain our entire
solar system. This density is still twenty-five times less than
the critical density of five atoms per cubic meter. Though it

may seem to imply perpetual expansion by a wide margin, the actual situation is less straightforward due to the mysterious dark matter. As explained in chapter 5, galaxies, and even entire clusters of galaxies, would fly apart unless they were held together by the gravitational pull of five to ten times more "dark" material than we can actually see. We think dark matter is probably some kind of particle, with no electric charge, left over from the early universe, but, whatever it is, it is important for the fate of our universe.

The Greek letter omega (Ω) is used to denote the average density divided by the critical density. If Ω exceeded 1, gravity could lead to eventual recollapse. According to our current inventory, ordinary atoms contribute only 4 percent of the critical density—in other words, if there were nothing else in the universe, Ω would be only 0.04. About half of these atoms make up the stars and gas within galaxies; the rest are mainly in intergalactic gas. Dark matter contributes five to ten times more, but that is still no more than about 0.3 of the density needed to halt the expansion. This finding suggests that the universe is not slowing down enough to ever come to a halt.

Acceleration?

There is another approach to this long-range cosmic forecasting that at first sight seems more straightforward: namely, to look directly for the difference between the expansion rate a few billion years ago and the present rate, and then extrapolate the trend forward. This comparison is possible in principle because the redshifts of distant objects tell us how they were moving when their light set out, not how they are moving now. The practical problem is that this technique requires objects bright enough to be detectable far away and, more demandingly, standardized enough that their apparent brightness tells us their distance. Galaxies themselves are not suitable. They are a rather varied zoo of objects, and it is uncertain how their brightness changes over time: their constituent stars age and die, new stars form, and so forth.

The best standard candle yet recognized is a type of supernova that is triggered by a nuclear explosion. These so-called Type 1A supernovae are, in effect, thermonuclear bombs—exploding stars with a standardized yield. The first results from studies of distant supernovae, announced in 1998 by two international teams of researchers, caused a stir. There certainly did not seem to be as much deceleration as one would expect in the simplest kind of universe where Ω was exactly one. That in itself was not surprising because it already seemed clear that there was not enough dark matter to make Ω more than about 0.3. But it *was* a surprise that the expansion seemed actually to be *speeding up*.

Science magazine rated this finding as the No. 1 scientific discovery of 1998, in any field of research, but this accolade may have been premature. The observations that were made are right at the limits of what is possible with existing telescopes. One still worries about possible complications and errors. For instance, the supernova "bombs" going off in the remote past, and observed as faraway objects with large redshifts, may not have been quite the same as those going off now, nearby. Or a fog due to intergalactic dust might make distant supernovae appear fainter than they actually are. The case is not yet overwhelming, but every month more supernovae are added to the sample, and within two or three years the various loopholes should be closed. It is fortunate that there are still two teams working independently, since the long-term prognosis for the universe is too important to be decided without a second opinion.

Acceleration implies an extra cosmic force—some kind of cosmic repulsion that overwhelms gravity. This idea is not in itself new: it goes back to Einstein in 1917. At that time, astronomers only really knew about our own Galaxy. Not until the 1920s did a consensus develop that Andromeda and similar "spiral nebulae" were actually separate galaxies, each comparable to our own. It was natural for Einstein to presume that the universe was static, neither expanding nor contracting. He found that a universe could not persist in a static state unless gravity was counteracted by an extra force. He incorporated an extra number into his equations, which he called the "cosmological constant," denoted by the Greek letter lambda (λ). This introduced a repulsive force—a kind of "antigravity"—which counterbalanced the ordinary gravity and allowed a static universe that was finite but unbounded. In Einstein's static universe, any light beam you transmitted would return and hit the back of your head.

Einstein, in his later life, rated lambda as his "biggest blunder," because if he hadn't introduced it to permit a static universe he might have predicted the expansion before Edwin Hubble's discovery in 1929. Einstein's motive for inventing lambda has been obsolete for seventy years; but that does not discredit the concept itself. On the contrary, lambda now seems less ad hoc than Einstein thought it was. Lambda can be envisioned as energy that is somehow contained even in empty space (known as "vacuum energy"). According to our present concepts, empty space is anything but simple. All kinds of particles are latent in it, and on an even tinier scale, it may be a seething tangle of strings. From our modern perspective, the puzzle is not why there should be a lambda, but why it is not much, much higher.

If there is energy in empty space (equivalent, as Einstein taught us, to mass through his famous equation $E = mc^2$), why does it have the opposite effect on the cosmic expansion from atoms, radiation, and dark matter, all of which tend to slow down the expansion? The answer depends on a feature of Einstein's theory that is far from intuitive: gravity, according to the equations of general relativity, depends not just on energy (and mass), but on pressure as well. A generic feature of the vacuum is that if its energy is positive, then its pressure is negative—in other words, it has a "tension," like stretched elastic. The net effect of vacuum energy is then to accelerate the cosmic expansion. It has a huge negative pressure and so, according to Einstein's equations, it pushes rather than pulls.[1]

If lambda represents the energy latent in space, which we realize has intricate structure on subatomic scales, the best theoretical guess is that it should induce a cosmic repulsion *120 powers of 10 stronger* than is actually claimed. There is a widely popular idea called "inflation" (discussed in chapter 9) which postulates that, very early in cosmic history, there was indeed a repulsion as fierce as this. If so, how could it have switched off (or somehow be neutralized) with such amazing precision?

Most physicists suspected that some process, not yet understood, made the resultant vacuum energy *exactly* zero, just as other features of our universe—for instance, its net electric charge—are believed to be. Recent research shows that it is not zero, but it is still very, very, small. Why? There clearly has to be some impressive cancelation, but why should this be so precise that it leads to a row of 119 zeros after the decimal point, but not 120 or more?

A slightly different theory speculates that the repulsion is not actually due to empty space (the vacuum), but that there is some all-pervasive fluid that has negative pressure and therefore exerts a gravitational repulsion, but which dilutes and decays during the expansion, so that by now it is guaranteed to be very small. This mysterious substance—a kind of dark energy—has been dubbed "quintessence."[2]

Evidence for "Flatness": Concordant Measures of Our Universe

Other data, quite independent of the supernova evidence for an accelerating universe, now support the case for lambda or quintessence. These data come from precise measurements of the "afterglow" of the Big Bang. This radiation is not completely uniform across the sky: there is a slight patchiness in the temperature, caused by the ripples that evolve into galaxies and clusters. Theory tells us that there's a certain wavelength on which the fluctuations—acoustic vibrations of the universe, as it were—would be maximal. How large these dominant ripples appear on the sky—whether, for instance, they are one degree across or only half a degree—depends on the geometry of the universe, which in turn depends on the mass and energy that it contains.

At almost any cosmology conference, one or more speakers are sure to project a picture resembling figure 7.2, which depicts the amplitude of the various vibration modes of the early universe: small angles to the right, large ones (lower harmonics) to the left. Maps of the background temperature, made from high, dry mountain sites, from Antarctica, or from long-duration balloon flights, have pinned down the angular size of these dominant ripples with better than 10 percent precision. The results indicate a "flat" universe: the relation between distance and angles is the same as in Euclidean space.

This is what would have been expected if dark matter provided the full critical density—if, in other words, Ω had been 1. On the other hand, a value of 0.3 for Ω would imply

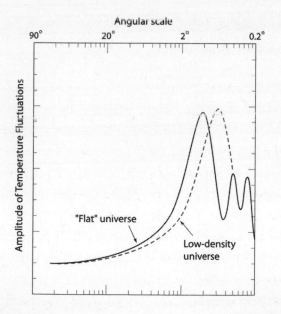

7.2
Predicted fluctuations in the background temperature in a flat universe. In a low-density open universe, the peak would occur on a smaller angular scale.

(if there were no other energy in the universe) an angle smaller by a factor of 2—definitely in conflict with observations. However, if vacuum energy (lambda) or quintessence contributes, in effect, the other missing 0.7, we get consistency with the data. Even these exotic forms of energy can modify the geometry of the universe and render it flat. However, because they contribute a negative pressure (despite having positive energy), they exert an antigravity or cosmic repulsion and therefore can account for the supernova evidence for an accelerating expansion.

Between 1998 and 2000, a remarkable concordance emerged in the results from several independent methods of measuring the key numbers describing our universe. It seems that the universe is flat, in harmony with theoretical prejudices. But its contents are an arbitrary-seeming mix of strange ingredients. Ordinary atoms (baryons) in stars, nebulae, and diffuse intergalactic gas provide just 4 percent of the mass-energy; dark matter provides about 30 percent; and dark energy the rest (i.e., about 66 percent). The expansion accelerates because dark energy (with negative pressure) is the dominant constituent. Of the atoms in the universe, only about one-half are in galaxies and the rest are diffusely spread through intergalactic space. The most conspicuous things in the cosmos, the stars and the glowing gas in galaxies, are less than 2 percent of the universe's total budget of mass-energy, an extraordinary turnaround from what would have been the natural presumption at the start of the twentieth century.[3]

In still earlier centuries, the classical view was that everything in the "sublunary sphere" consisted of the four "elements"—earth, air, fire and water—but the heavens were constituted from some quite different "fifth essence." This concept was laid to rest in the nineteenth century, when studies of stellar spectra showed, as described in chapter 3, that stars were made of the same stuff as Earth. But modern cosmology revives a similar antithesis. It looks as though the mysterious "antigravity substance"—vacuum energy, or quintessence—provides the dominant mass-energy in our universe even though it plays no role in stars or galaxies. It could be a vestige of the (far fiercer) force that, according to the theory of "inflation," drove the rapid expansion in the early universe, though we still face the question of why the force, once so strong, is now so feeble. Its nature is a challenge to theorists: it holds important clues to the early universe, and to the nature of space.

8   The Long-Range Future

A Darker Future

Five billion years from now, when the Sun dies, the galaxies will be more widely dispersed and intrinsically somewhat fainter because their stellar population will have aged, and less gas will survive to form bright new stars. But what might happen still farther ahead? We can't predict what role life will eventually carve out for itself: it could become extinct, or it could achieve such dominance that it can influence the entire cosmos. The latter is the province of science fiction, but it can't be dismissed as absurd. After all, it has taken little more than one billion years for natural selection to lead from the first multicellular organisms to Earth's present biosphere, which includes us. By the time the Sun dies, five times longer will have elapsed—time enough for Earth's biosphere (and any others that exist) to be inconceivably transformed, even if future species emerged on the timescale of biological natural selection. Future changes would occur faster still if they are artificially directed—on a cultural or historical timescale.

We can, despite these inponderables, offer tentative long-range projections for the gross features of our universe. It seems fated to continue expanding, and even—if current evidence is borne out—accelerating. Eventually, even the slowest-burning stars will die, and all the galaxies in our Local Group—our Milky Way, Andromeda, and dozens of smaller galaxies—will merge into a single system. Most of the original gas would by then be tied up in the dead remnants of stars; some would be black holes, others would be cold neutron stars or white dwarfs.

Still farther ahead, events far too rare to be discernible today, such as stellar collisions, could come into their own. Stars are so thinly spread through space that collisions between them are immensely infrequent (fortunately for our Sun), but their number would mount up. The terminal phases of galaxies would be sporadically lit up by intense flares, each signaling an impact between two dead stars. Another very slow process is gravitational radiation, the weak ripples in space that are generated by any heavy object that moves and changes its shape. These waves carry away energy. Their effects are imperceptible today, except in a few binary stars where the orbits are specially close and fast; but they would, given enough time, grind down all stellar and planetary orbits.

Eventually, even black holes will decay. The surface of a black hole is made slightly fuzzy by quantum effects, and consequently it radiates. This effect would be important in our present universe if it contained miniholes the size of atoms: such black holes would erode away, emitting radiation and particles; the smaller they get, the more powerful and energetic the radiation would be, and they would eventually disappear in an explosion. But it seems unlikely that such miniholes exist. They could form only in the hyperdense early universe, and even there this would be unlikely unless conditions were far more turbulent than theory suggests. To form them today, a kilometer-sized asteroid (or something of similar mass) would need to be compressed to the size of an atomic nucleus.

The evaporation of black holes, being a quantum process, is far less important for big holes: the time it would take for a hole to erode away depends on the cube of its mass. The lifetime of a hole with the mass of a star is 10^{66} years. Even black holes as heavy as a billion suns, such as those that lurk in the centers of galaxies, would have eroded away before 10^{100} years had elapsed.

The Asymptotic Future of Life

Cosmologists have produced an immense speculative literature on the ultra-early universe. In contrast, cosmic futurology has been left to science fiction writers. I myself can claim to have made one of the first scientific contributions, back in

1968, when I wrote a short paper entitled "The Collapse of the Universe: An Eschatological Study." Many cosmologists then suspected that we might live in a bounded universe that would end in a Big Crunch, and I calculated what might happen after the cosmic expansion had come to a halt and the universe had started to recollapse. During the countdown to the crunch, galaxies would merge together, and individual stars would accelerate to almost the speed of light (rather as the atoms speed up in a gas that is compressed). Eventually, these stars would be destroyed because the heat irradiating their surfaces (the blue-shifted radiation from other stars) would be hotter than their interiors.

Ten years later, Freeman Dyson made the subject scientifically respectable: he published, in *Reviews of Modern Physics* (an austere and prestigious academic journal), a fascinating and detailed article called "Time without End: Physics and Biology in an Open Universe." The evidence for an ever-expanding universe was then less clear than it is now. But Dyson already had his prejudices: he would not countenance the Big Crunch option because it "gave him a feeling of claustrophobia." He discussed the prognosis for intelligent life. Even after stars have died, he asked, can life survive forever without intellectual burnout? Energy reserves are finite, and at first sight this might seem to be a basic restriction. But he showed that this constraint was actually not fatal. As the universe expands and cools, lower-energy quanta of energy (or, equivalently, radiation at longer and longer wavelengths) can be used to store or transmit information. Just as an infinite series can have a finite sum (for instance, $1 + 1/2 + 1/4 + \ldots = 2$), so there is no limit to the amount of information that could be processed with a finite expenditure of energy. Any conceivable form of life would have to stay ever cooler, think slowly, and hibernate for ever longer periods. But there would be time to think every thought. As Woody Allen once said, "Eternity is very long, especially toward the end."

Dyson imagined the endgame being spun out for a number of years so large that to write it down you would need as many zeros as there are atoms in all the galaxies we can see. At the end of that time, any stars would have tunneled into black holes which would then evaporate, in a time that is in comparison almost instantaneous.

In the twenty years since Dyson's article appeared, our perspective has changed in two ways, and both make the outlook more dismal. First, most physicists now suspect that atoms don't live forever. In consequence, white dwarfs and neutron stars will erode away, maybe in 10^{35} years; the heat generated by atomic decay inside dead stars makes them glow, but as dimly as a portable heater. By then our Local Group of galaxies would be just a swarm of dark matter and some electrons and positrons. Thoughts and memories would only survive beyond the first 10^{35} years if downloaded into complicated circuits and magnetic fields in clouds of electrons and positrons—maybe something that would resemble the threatening alien intelligence in *The Black Cloud*, the first and most imaginative of Fred Hoyle's science fiction novels, written in the 1950s.

Dyson was optimistic about the potentiality of an open universe because there seemed to be no limit to the scale of artifacts that could eventually be constructed. He envisioned

THE LONG-RANGE FUTURE

the observable universe getting ever vaster; moreover, many galaxies, whose light has not yet had time to reach us, would eventually come into view and therefore within range of possible communication and "networking." Gravity tends to slow the recession of distance galaxies even though they move ever farther away, so the disruptive effect of the expansion becomes less important. As described in chapter 7, cosmologists suspect that the expansion is not slowing down: some repulsive force or antigravity seems to be pushing galaxies apart at an accelerating rate. The long-term future is then more constricted. Galaxies will fade from view even faster: they will get more and more redshifted—their clocks, as viewed by us, will seem to run slower and slower and freeze at a definite instant, so that even though they never finally disappear, we would see only a finite stretch of their future. The situation is analogous to what would happen to Star Trekkers who fell into a black hole: from a vantage point safely outside the hole, we would see our infalling colleagues freeze at a particular time, even though they experience, beyond the horizon, a future that is unobservable to us.

Our own Milky Way Galaxy, Andromeda, and the few dozen small satellite galaxies that are in gravitational grip of one or other of them will merge together into a single amorphous system of aging stars and dark matter. In an accelerating universe, everything else disappears beyond our horizon; if the acceleration is due to a fixed lambda, this horizon will never get much farther away than it is today. So there is a firm limit—though, of course, a colossally large one—to how large any network or artifact can ever become. This translates into a definite limit on how complex anything can get. The inherent "graininess" of space sets a limit to the intricacy that can be woven into a universe of fixed size.[1] The best hope of staving off boredom in such a universe would be to construct a time machine and, subjectively at least, exhaust all potentialities by repeatedly traversing a closed time loop.

These long-range projections involve fascinating physics, most of which is quite well understood. But readers who are really concerned about what happens in zillions of years should be mindful of some uncertainties. First, we can not be absolutely sure that the regions beyond our present horizon are like the parts of the universe we see. On the ocean, there could be something extraordinary just beyond the horizon. Likewise, a universe that's decelerating, though seemingly not enough ever to stop, could eventually be crushed by denser material not yet in view. Even if that did not happen, the trend toward greater smoothness on larger scales may not continue indefinitely. There could be a new range of structures, offering new supplies of energy on scales far larger than the part of the universe that we have so far observed.

Second, we don't know what may eventually happen to the quintessence—the mysterious energy in space that drives the accelerating cosmic expansion. This residual energy could convert into some new kinds of particles. If this conversion happened smoothly, it would not lift the terminal gloom. However, the residual energy could decay in bubbles whose surfaces crash together, giving rise to energy concentrations where atoms might possibly even be regenerated. This revived universe would be patchy, with islands of renewed activity separated by vast voids. To quote Paul Steinhardt, would future beings "figure out that their origin was the isotropic universe we see around us today? Would they ever know that the universe had once been alive and then died, only to be given a second chance?"[2]

A more disconcerting prospect is that empty space could be vulnerable to a catastrophic transfiguration. Very pure water can "supercool" below its freezing point, but it will suddenly freeze when a speck of dust is put into it. In an analogous way, our present "vacuum" may be merely meta-stable and could then transform to a quite different universe, governed by different laws, perhaps with a large negative lambda that would cause everything to implode rather than accelerate outwards. It is a cosmic version of what happens to the world's water in Kurt Vonnegut's *Ice Nine*: some event triggers formation of an anomalous form of ice, solid at or-dinary temperatures, which then catalyzes the conversion of all the water on Earth into this fatal new state.

There are periodic scares that this kind of transition could be induced artificially by high-energy particle colli-sions in accelerator experiments. It is reassuring, however, that far more energetic collisions—involving cosmic ray par-ticles—have been happening naturally for billions of years, without tearing the fabric of space.

Another route to Armageddon could be the conversion of ordinary atomic nuclei into hypothetical particles called strangelets, which contain a third kind of quark, additional to the kinds that make up normal protons and neutrons. If strangelets can exist as stable objects, and if their electric charge is negative (which is thought very unlikely), they could attract other nuclei and then, by contagion, convert their surroundings, and eventually the entire Earth, into so-called strange matter. This eventuality was taken seriously enough by those in charge of the Brookhaven accelerator that they commissioned an expert assessment on whether experiments that crash together very heavy nuclei could trigger such a catastrophe.[3]

The experts offered reassurance. However, I don't think we can sleep completely soundly unless we are sure that essentially identical events have already happened naturally without any disaster ensuing. The acceptable level of risk in such an experiment is surely less than one chance in a trillion. Theoretical arguments alone cannot offer adequate comfort at this level: only a recklessly overconfident theorist would stake a bet on the validity of his assumptions at odds of a trillion to one.

Our remote descendants are likely to have an eternal future (unless they come within the clutches of a black hole). Nonetheless, it is worth noting that the alternative fate—being snuffed out in a Big Crunch—could be an enriching experience. We have seen that events happen ever more slowly in the ever-expanding universe; the total number of discrete events or "thoughts" could then be bounded, even in an infinite future. John Barrow and Frank Tipler have emphasized that, in a collapsing universe, the converse is possible: there can be an infinite number of "happenings" within a finite time. Cosmologists are used to the idea that a lot could happen in the initial instants after the Big Bang: as we extrapolate back to more extreme densities, time must be measured by a series of progressively smaller, more robust, and faster-ticking clocks. An infinite set of numbers can add up to a finite sum. Likewise, in the final instants before the crunch, we could not only see our entire earlier life flash by, but we could experience an infinite number of new events.

So much for the far future. Let's now go back to the beginning.

9 How Things Began: The First Millisecond

Tracing the Causal Chain Backwards: Confidence or Caution?

Astronomy and cosmology have a high profile and a positive image, in contrast to the ambivalence with which the public perceives, for instance, genetics or nuclear science. The essence of the new discoveries and concepts in these cosmic sciences can be conveyed, free of technicalities, to a wide general audience. I am uneasy, however, about the media's portrayal of the subject: all too often, claims are hyped up, only to be retracted later, with the retraction usually being more subdued. When this happens, it is often the scientists themselves (or their press offices) who are at fault, not the journalists: indeed, journalists may soon become as skeptical of some researchers as they routinely are of politicians. Exaggerated claims about cosmology plainly are not as damaging as, for instance, misleading reports about medical research that raise false hopes of miracle cures. But it is in the cosmologists' own interest to avoid hype. If we claim too often to make breakthroughs that overthrow all previous ideas, we'll surely lose even the credibility that we deserve. Cosmologists thus should not blur the distinction between what is well established and what is conjectural. Otherwise there are two downsides. On the one hand, flaky ideas may gain undue credence: on the other hand, robust skeptics, realizing that parts of the subject are indeed still speculative, may fail to appreciate that other parts are firmly battle tested.

We can now set our entire solar system in a grand, evolving scenario stretching back to a Big Bang—an era when everything was hotter than the centers of stars and expanding on a timescale of a few seconds. The evidence is as compelling as most of the claims made by geologists and paleontologists about the history of Earth—indeed, it is a lot more quantitative. In chapter 5, I sketched how an expanding fireball, fluctuating slightly from place to place, can end up, 10 billion or more years later, resembling our present cosmos. Computer simulations can mimic how this happens.

Galaxies of stars, grouped into clusters, emerge mainly through the action of gravity—the same force that Newton realized held the planets in their orbits. One can extend the parallel with Newton. He knew why planets traced out ellipses, but the initial "setup" of the solar system was a mystery to him: Why were the orbits of the planets all close to a single plane, the ecliptic, while the comets came from random directions? In the second edition of *Opticks* (1717) he wrote: "Blind fate could never make all the planets move one and the same way in orbits concentrick. . . . Such a wonderful uniformity in the planetary system" must, he claimed, be the result of providence. This coplanarity of the orbits, however, is now understood: it is a natural outcome of the solar system's origin as a spinning protostellar disk. Indeed, we can trace things back farther still—to the first second of the Big Bang.

Conceptually we face a similar barrier to Newton. He wondered why the planets moved in a particular plane. We wonder why the few-second-old universe was set up to expand at a particular rate and with a particular set of ingredients.

In another respect, too—in the interface between cosmology and religion—there seems to have been no qualitative change since Newton's time. His discoveries triggered a range of religious (and antireligious) responses. Likewise, modern cosmologists evince a variety of religious attitudes: there are both traditional believers as well as hardline atheists among them. My personal view—a boring one for those who wish to promote constructive dialogue (or even just unconstructive debate) between science and religion—is that, if we learn anything from the pursuit of science, it is that even things as "elementary" as atoms are quite hard to understand. This should induce skepticism about any dogma, or about any claim to have achieved more than a very incomplete and metaphorical insight into any profound aspect of our existence. It is of interest, however, that E. O. Wilson, in his book *Consilience*, regarded deism (if not theism in general) as "a problem in astrophysics."

Back into the First Millisecond

The recipe that describes our one-second-old universe is simple enough to be encapsulated very briefly:

1. The proportions of ordinary atoms, dark matter, and radiation
2. The cosmic expansion rate
3. How smooth the expansion is—essentially the value of just one number, Q, which determines the texture and scale of structures
4. The basic properties of atoms and their nuclei.

This recipe must be the outcome of what happened still earlier, within the first tiny fraction of a second. I emphasized that we could confidently trace things back to the first few seconds of cosmic expansion. The background radiation, and the helium and deuterium, are well-measured fossils of that era. At even earlier times, conditions become more extreme and unfamiliar. For the first 10^{-5} seconds, the material would be denser than an atomic nucleus; protons and neutrons would, in effect, split into their constituent quarks. Experimenters at the CERN laboratory in Geneva and at Brookhaven National Laboratory in the United States have replicated these conditions on a tiny scale by crashing together nuclei of lead and gold accelerated to almost the speed of light.[1] Farther back still, the energies and densities are so extreme that we have even less foothold in experiment.

In this context, we should think of time on a logarithmic scale. Each factor of 10—each extra zero after the decimal point on the cosmic clock—should count equally. Ideas about this *ultra*-early era are still tentative—they should be preceded by the sort of health warning printed on cigarette packets—but nonetheless there has been immense progress in pushing back the frontier.

An absolute limit to any credible backward extrapolation is set by quantum theory. The key concept of this theory is Heisenberg's uncertainty relation, which tells us that the more accurately you want to locate or localize something, the more energetic are the quanta—the packets of energy—you need. There is a limit when the energy is so concentrated that it risks imploding into a black hole. This limit is the *Planck length*: its value is 10^{-33} cm—smaller than a proton by about 19 powers of 10 (see the diagram in the Appendix). This minuscule length, divided by the speed of light, defines the smallest measurable time interval, the *Planck time*, about 10^{-44} seconds.

In the first trillionth of a second, each particle would have carried more energy than the most powerful accelerators can attain. But a trillionth of a second exceeds the Planck time by more than 30 powers of 10; therefore, there are many decades of logarithmic time during which the microphysics is uncertain because of the high energies and densities, but where conditions are nonetheless not so extreme that we need to worry about quantum gravity and the (perhaps) discrete structure of space and time.

Some of the formative processes in our universe are believed to have happened when it was about a hundred million times older than the Planck time, but still only 10^{-36} seconds old. In particular, this could have been when the present mix of atoms and radiation was imprinted. As mentioned earlier, the most abundant entities in the universe are the quanta of radiation—photons—that make up the residual heat from the Big Bang. Atoms, which are crucial to our existence and to the existence of all planets, stars, and galaxies, are 2 billion times less abundant than these photons. But the surprising thing, the theoretical conundrum, is why there isn't just radiation—why are there any atoms at all?

If you were setting up a universe in the simplest form, you might mix in equal amounts of matter and antimatter: as many antiprotons as protons; as many antiquarks as quarks. But in such a universe, all the matter and antimatter would annihilate into radiation.

What is it that allowed some matter, but not antimatter, to survive? The answer depends on the ideas pioneered by the great Russian physicist and one-time dissident Andrei Sakharov, combined with experimental evidence that matter and antimatter are not precise mirror images of each other. This asymmetry slightly favors matter over antimatter very early in cosmic expansion. The favoritism is only about one part in a billion, though. But that small effect is crucial: we owe our very existence to a difference in the ninth decimal place. For every billion pairs that annihilate, eventually yielding the photons that make up the microwave background, one extra unpaired quark is left over.

There is an important contrast here with electric charge. Our universe has zero charge, where any positive charged particle is exactly balanced by a negatively charged particle. But the baryon number—the number of protons minus the number of antiprotons—is not absolutely conserved. That is why the excess can be imprinted in the early universe. There is a price to pay, however. Since the number of protons is not strictly conserved, they can decay spontaneously, even if there is no antiparticle to annihilate with. The difficulties our descendants will face in 10^{35} years, when their constituent atoms gradually erode away, are the payback or downside for the good fortune that favored matter over antimatter in the first place. This prevented everything from annihilating into gamma rays before galaxies, and their constituent stars, had a chance to form.

The common presumption is that a consistent favoritism for matter over antimatter prevails throughout the observable universe, but what is the hard evidence for this? Antimatter can survive only if it is well quarantined from ordinary matter. Otherwise it annihilates, and the entire mass-energy (Einstein's mc^2) is transformed into radiation, mainly gamma rays. Our own Galaxy could certainly not have started off as an equal mix of matter and antimatter: if it had, nothing would now be left, because its contents have been churned up and mixed by the turbulent processes attendant on the births and explosive deaths of stars. Even in entire clusters of galaxies, there is enough mixing that we would expect stronger gamma-ray emission than is actually detected. On the scale of the largest superclusters, we cannot be so dogmatic. Alternating domains of matter and antimatter could be spread through the universe, provided these were on scales larger than a hundred million light years.

Why Expansion?

We have an explanation of why our expanding universe contains a mix of radiation and matter. But there is a still more fundamental question—how did it begin, and why is it expanding the way it is?

It is seriously misleading to think of the Big Bang as being triggered by an explosion. Bombs on Earth, or supernovae in the cosmos, explode because a sudden boost in internal pressure flings the ejecta into a low-pressure environment. But in the early universe the pressure was the same everywhere: there was no empty outside region. The extra gravity due to the pressure and heat energy actually helps to slow down the expansion. We need some other explanation for what banged and why it banged. Furthermore, there is something very special about the expansion rate, as illustrated in figure 9.1.

Suppose you were setting up an expanding universe. The trajectory it would follow would depend on the impetus it was given. If it were too fast, then the expansion energy would, early on, have become so dominant that galaxies and stars would never have been able to pull themselves together via gravity, and then condense out. On the other hand, if the initial impetus were insufficient, a premature Big Crunch would quench evolution when it had barely begun. Back at one second (remember that we have good grounds for extrapolating back that far) its kinetic and gravitational energies must have differed by less than one part in a million billion (one in 10^{15}).*

When this problem was first recognized, there were two

---

*If the postulated starting point were much earlier, then the inferred precision would have been still greater; far larger numbers, indicating far more impressive-seeming fine-tuning, are quoted by other authors. I have taken one second as the starting point because we have firm evidence (from helium and deuterium) about the physical conditions and expansion rate at that time, whereas at earlier times the evidence is less direct.

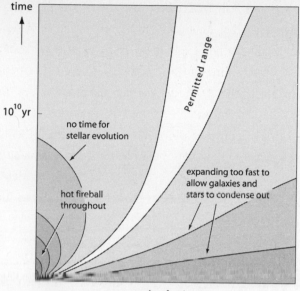

time

10^{10}yr

Permitted range

no time for
stellar evolution

hot fireball
throughout

expanding too fast to
allow galaxies and
stars to condense out

scale of universe

9.1

Trajectories for nonhabitable expanding universes. If the expansion is
too slow, recollapse occurs before there has been a chance for complex evolution. If it is too fast, gravity cannot pull structures together,
and no galaxies or stars would form.

responses. One was that the universe would probably sometime recollapse, but that we should not be surprised that it
had heaved itself up to such a vast size. If it hadn't, we would
not have had time to evolve before the collapse and crunch,
and we would not be here to wonder about it. (Arguments of
this kind will recur in chapter 11). The second response,
which has been the dominant one for the last twenty years,
has been to seek some reason for the close balance between
expansion energy and gravity.

Why is our universe so big? Why, indeed, is it expanding

at all? By far the most plausible answers involve a so-called inflationary phase. According to this concept, very early on, when everything astronomers now see was literally of microscopic size, the dominant stuff in the universe was not ordinary particles or radiation, but concentrated dark energy latent in the original space. The nonintuitive aspect of this dark energy is that it can exert a negative gravity—it pushes rather than pulls. Like a very high lambda, the "dark energy" overwhelmed ordinary gravity, generating an immensely powerful antigravity, or cosmic repulsion. The expansion was exponential; the scale doubled, then doubled, and then doubled again. And then, the fierce repulsion switched off; some of the dark energy was converted into ordinary energy, generating the heat of the primeval fireball and initiating the more familiar expansion process that has led to our present habitat.

Within about 10^{-36} seconds—a trillionth of a trillionth of a trillionth of a second—a microscopic patch could have inflated large enough to encompass everything we now see, and to establish the fine-tuned balance between gravitational and kinetic energy.

It may take a very long time to stop exponential expansion once it has started. Theorists refer to this situation as the "graceful exit" problem. Our universe will end up being stretched flat, rather as any part of a wrinkled surface becomes smooth if it is stretched enough. On a two-dimensional surface, the effects of curvature show up in the geometry: for example, on Earth's curved surface, the three angles of a triangle add up to more than 180 degrees. Likewise, the geometry of three-dimensional space depends on its curvature, but inflation theories predict that our universe has the special property of "flatness." In a flat universe there is a definite relation between the angular size of remote objects and their distance from us and measurements of the background radiation confirm this prediction.[2]

The mechanism that drove inflation is, in essence, the same as the one that causes acceleration at the present day, except that the repulsive force—and the energy and tension in space—was higher by 120 powers of 10.

Inflation, once started, is likely to overshoot, leading to a flattened domain extending much farther than the 10 billion light-year dimensions of our observable universe. The distance to the "edge" could be a number with millions of zeros. The 60 powers of 10 difference between the Planck length and our horizon scale would then be nothing compared with the leap beyond to the real limit of our universe. In this expanse of space, far beyond the horizon of our observations, the combinatorial possibilities are so immense that close replicas of our Earth and biosphere would surely exist, however improbable life itself may be. Indeed, in a sufficiently colossal cosmos there would, somewhere, be exact replicas not just of our Earth, but of the entire domain (containing billions of galaxies, each with billions of stars) that lies within range of our telescopes.

Furthermore, even this stupendous expanse of space may not be everything there is. Some theories suggest that our Big Bang was not the only one. Space inflates so fast that it opens up enough "room" to trigger a perpetual succession of independent Big Bangs. This line of speculation dramatically enlarges our concept of reality—from a universe to a multiverse.

As I shall discuss in chapter 11, this scenaro might make some features of our own cosmic habitat less surprising.

Most theorists regard inflation as a beautiful idea, which they will cling to until something better comes along. Indeed, there are intimations that something might, that extra spatial dimensions beyond the usual three may lead us to another paradigm. Some have never been enraptured by the inflation theory. Roger Penrose, for instance, thinks inflation is a "fashion the high energy physicists have visited on the cosmologists"; he notes that "even aardvarks think their offspring are beautiful." He believes that an understanding of our Big Bang must await a new conceptual breakthrough that will transform our concepts of quantum theory as well as of gravity; he also—and here he is farther out on a limb—suspects that this same breakthrough will illuminate the enigma of consciousness as well.

Everyone accepts that new insights—or some firmer way of deciding among the various ideas already proposed—will be needed before we can describe the ultra-early universe with any confidence. Some features of our present universe may be sensitive to the details of those ideas; observations may therefore help to select among rival theories. For instance, inflation suggests an origin for the ripples that show up as nonuniformities in the background temperature and are the embryos of galaxies and clusters of galaxies. They are quantum vibrations, generated on a scale that is actually microscopic, that have inflated so much that they now stretch across the sky in an amazing link between cosmos and microworld. Related to these ripples would be gravitational waves—oscillations in the fabric of space itself, crisscrossing the universe at the speed of light. These waves make everything shake slightly. The effect is minuscule and its detection poses a formidable technical challenge, even for large experiments in space that are being designed to search for gravitational waves. Such waves may manifest themselves indirectly, however: they cause a small polarization in the microwave background radiation, and if this polarization could be measured, we would have a direct diagnostic of what happened during inflation.

We can make various guesses about the physics, calculate what each guess implies for the microwave background fluctuations, the way galaxies cluster, the gravitational waves, and so on, and thereby at least narrow down the options. Astronomers normally just apply laboratory physics (except where gravity is concerned). Perhaps they can now return the compliment by learning things about "extreme physics" that cannot be checked in the laboratory.

Cosmological Research Ten Years from Now

I would bet reasonable odds that by the year 2010 we will be very confident of what the dominant dark matter is, the value of Ω, and the properties of the dark energy in the vacuum. If that happens, it will signal a great triumph for cosmology: we will have taken the measure of our universe, just as, over the last few centuries, we have learned the size and shape of our Earth and Sun. And, subject to some provisos mentioned in the next chapter, we will know the long-range cosmic forecast.

The challenges after 2010, will then be of two very different types. That's because cosmology has two faces: it is a fundamental science, but it is also the grandest of the environmental sciences. The Canadian theorist Werner Israel has likened this dichotomy to the contrast between chess and mud wrestling. The community of cosmologists is, perhaps, such an ill-assorted mix of extreme refinement and extreme brutishness (only in intellectual style, of course).

A decade or so from now, those of us who are happier wallowing in the mud—and more intellectually fitted for that than for chess—will be enthusing about ever-more-detailed observations from telescopes both on the ground and in space. Massive computer simulations will enhance our understanding and intuition of how galaxies, stars, and planets form.

The data rate will be so colossal that the entire process of analysis and discovery will be automated. Astronomers will focus attention on heavily processed population statistics for planets, stars, and galaxies, and on the best cases of each phenomenon—for instance, the most Earthlike planets and the pathological objects that may hold exciting clues to extreme physics. Dramatic advances in computer technology will allow virtual-reality experiments on stellar collisions, black holes, and other phenomena. Astronomers will be part of a larger and more widely dispersed community than they are today. Technology will allow more democratic access to data that in the past were accessible only to an elite (or otherwise favored) minority. Detailed maps of the sky will be available to anyone who can access or download them from the Internet. There will be virtual observatories. Enthusiasts anywhere in the world will be able to participate in exploring our cosmic habitat, checking their own hunches, seeking new patterns, and discovering unusual objects. There will be armchair observers as will as armchair theorists.

But the "chess players" will still, I would guess, be seeking a deeper understanding of the very beginning. The search for unified theories of the cosmos and microworld will not be exhausted (though it may have exhausted the searchers). In the remaining two chapters, I will speculate on the scope and limits of such insights.

PART III

Fundamentals and Conjectures

10 Cosmos and Microworld

From Nothing?

Everything astronomers can see, stretching out to distances of 10 billion light-years, emerged from an infinitesimal speck. This astonishing idea becomes easier to swallow if we realize that, in a sense, the universe's net energy can be zero. Everything has an energy equal to mc^2, according to Einstein's famous equation. But everything also has negative energy because of gravity. We on Earth have an energy deficit compared to an astronaut in space. But the deficit due to all the masses in the universe added together could amount to *minus mc^2*.

In other words, the universe makes for itself a gravitational pit so deep that everything in it has a negative gravitational energy that exactly compensates for its rest-mass energy. So the energy cost of inflating our universe could actually be zero.

Cosmologists sometimes claim that the universe can arise "from nothing." But this is loose language. Even if shrunk to a point or a quantum state, our universe is latent with particles and forces: it has far more content and structure than what a philosopher calls "nothing."

I mentioned in chapter 9 that modern cosmologists still, like Newton, come up against a barrier when tracing the causal chain backward: at some stage, the only answer to be given is "things are as they are because they were as they were." We are far from having a unique, self-consistent picture of our universe. In one sense, cosmologists may actually be *worse off* than Newton was. His laws are *autonomous*—orbiting planets obey them, but planets don't react back on the laws and modify them. On the other hand, one cannot count on this disjunction for an entire universe: the cosmos may determine the local laws as well as be governed by them.

The nineteenth-century physicist and philosopher Ernst Mach suggested that objects derived their inertia from some kind of interaction with the rest of the universe. This idea, dignified with the term "Mach's principle," is central to the nature of rotation and spin. Newton himself pointed out that, if you spin a bucket of water, the shape of the water's surface does not depend on whether it is rotating relative to the bucket walls: rotation is measured relative to a more global frame of reference, the so-called inertial frame. Mach conjectured that this inertial frame was actually fixed by the average motion of everything in the universe.

This might at first sight seem a meaningless distinction, but it is not. It turns out that the inertial frame defined, for instance, by the steady swings of a Foucault pendulum, is at rest relative to the distant universe. But some have thought it at least conceivable that this might not be so. Indeed, Einstein's equations certainly allow rotating universes. (One of these universes was the solution discovered by Gödel, which also allowed time travel, but there are others that are less bizarre.) Is it just a contingency that our own universe looks the same in all directions and has no special axis? Or is there some deeper and more restrictive principle that rules out rotating universes?*

Inconstant Constants?

Our universe may not be rotating, but it is certainly expanding—and changing as it does so. Some have argued that it would actually be surprising if, in a changing universe, the physical laws were *un*changing. Back in 1937 the great physicist Paul Dirac gave a specific argument of this kind, which suggested that Newton's gravitational constant G might be decreasing as the universe aged. The change would amount to one part in 10 billion each year.

---

*This is the same kind of question that we encountered in chapter 6 with regard to time machines: Is there some law that precludes closed time-like lines?

Dirac's motivation was interesting.[1] He noted that gravitational and electric forces both obey the inverse square law. Therefore, the ratio of the strengths of the electrical and gravitational forces between, say, an electron and a proton is a fundamental number and exceedingly large: about 10^{39}. Dirac was surprised to find that the size of the observable universe (the Hubble radius) exceeds the size of a proton by a factor also around 10^{39}. He then estimated the number of atoms in the observable universe (this is an even rougher estimate) to be about 10^{78}, the square of 10^{39}. Being reluctant to treat the similarities as a coincidence, he conjectured that there must be some underlying linkage between these large numbers—specifically, he suggested that Newton's constant, G, might be changing with the age of the universe, so that the two numbers increased in step.*

An effect as large as Dirac envisioned can now be ruled out, even by evidence within our solar system. If Dirac were right, all planetary orbits would spiral gradually outward, at a calculable rate, as the Sun's grip weakened. The Earth, when it formed, would therefore have been closer to the Sun than it now is. Moreover, the Sun would have been more luminous: stronger gravity would have caused its outer layers to press down harder on the hot core, enhancing the power output from the nuclear furnace. These two effects—that the Earth was closer then than it is today and the Sun was hotter than it is now—would have made the

---

*The number quoted in chapter 3 is 10^{36} rather than 10^{39} because there I considered the gravitational force between two protons, rather than between an electron and a proton. The Hubble radius is essentially the time since the Big Bang multiplied by the speed of light.

oceans boil when the Earth was young, contrary to what geology tells us.

We now know that G is not weakening at even a hundredth of the rate that Dirac wanted. The best evidence comes from the careful tracking of space probes and from a binary system containing two neutron stars whose orbits around each other can be monitored very accurately.

Of all the cosmic forces, gravity is the one that most obviously links with the large-scale universe. But could the cosmic expansion somehow induce changes in atoms—the electric and nuclear forces within them, the mass and charge on each electron, and so forth? As mentioned in chapter 2, astronomers already realized 150 years ago that the stars were made of the same stuff as Earth. We know we do not live in a complelely anarchic cosmos, where the atoms, and the laws governing them, vary capriciously from star to star. But the light from very distant galaxies set out when our universe was much younger and more compressed. Could the microphysics have been different then, in ways that would show up when astronomers analyzed this ancient light?

The "bar codes" in the spectra from even the remotest galaxies convey the message that all atoms, everywhere, seem closely similar: remote atoms do not differ by more than one part in a million from those we can study in the laboratory. Neither the charges nor the masses of electrons have changed by more than this amount in the billions of years since the light set out from these galaxies.

Another strong constraint comes from a surprising quarter: the Oklo mine in Gabon, West Africa. On this site, uranium ore and water accumulated about two billion years ago and "went critical." The geological deposits contain the outcome of radioactive decays in the remote past. The rare element samarium is especially interesting, because its rate of decay would be substantially altered even if the electron charge had differed by only one part in a million from its present value. Careful analysis of the 2-billion-year-old Oklo deposits constrains any changes in the physics of atoms just as stringently as astronomical observations can.[2]

The most distant galaxies are so far away that their light set out when the universe was only about a tenth of its present age. But could there have been more rapid and drastic changes even earlier? Here the best evidence comes from the outcome of nuclear reactions that occurred in the first few minutes after the Big Bang. (see chapter 5). Changes of more than a few percent in either the gravitational force or in the properties of electrons would have altered the yield of helium and deuterium rendering their abundances inconsistent with observations.

These constraints are already quite stringent, but astronomers should keep refining them because some modern theories actually predict very slow changes. What have come to be called "grand unified theories" suggest that, at the extreme temperatures attained 10^{-36} seconds after the Big Bang, when inflation is thought to have occurred, electrical and nuclear forces would have had the same strength; these forces would have differentiated as the expanding universe cooled. But the expected adjustments would happen so early that the "constants" really would seem unchanging over the entire span of time that we can now probe. Other, more radical, ideas involve extra dimensions that could induce larger and later changes in the forces and other basic features of the microworld. There is therefore a strong incentive to improve the sensitivity of these searches; perhaps astronomers can thereby provide a defining test for radical new theories in microphysics.

Our intuitions about space and time break down on the very tiniest scales, and at the very earliest times. We cannot describe the so-called Planck scales without going a step beyond even a grand unified theory to a theory that incorporates gravity and reconciles it with the quantum principle. This is unfinished business, but it is likely to involve jettisoning commonsense concepts of continuous time and three-dimensional space.

Other Dimensions?

Universes could have short lives, be governed by basic forces with different strengths, or even contain a different "zoo" of fundamental particles. But could their number of space dimensions be different from our actual three? Such spaces are routine to mathematicans, but the physics in them is harder to grasp. In a two-dimensional world, complicated networks are impossible without the wires crossing each other, though this does not preclude elaborate patterns of waves that can pass through each other. Even in one dimension, there could be intricate complexity. In his classic science fiction novel *Star Maker*, Olaf Stapledon envisages, among many imaginative cosmoses, a "musical universe, where creatures appeared to one another as complex patterns and rhythms of tonal character. A creature's body was a more or less constant tonal pattern, [which] could traverse other living bodies in the pitch dimension much as wave-trains in a pond may cross one another."

Stapledon's musical universe has, in effect, just one spatial dimension, in addition to time. Our actual space-time has, of course, three spatial dimensions. Time, the fourth, is different from the other three insofar as it has an arrow: we seem to be dragged only one way in it (forward). Three-dimensional space has special features. For instance, if an object is rotated in an arbitrary way, it takes three numbers—the same as the number of dimensions—to specify the rotation: two to fix the direction of the rotation axis, and one to specify the angle through which it rotates around that axis. (In two dimensions, rotations are defined by just one number; in four dimensions, six are needed.)

It is actually because there are three space dimensions that electric and gravitational forces obey an inverse square law. This dependence is easiest to appreciate in terms of Faraday's concept of lines of force. A shell of radius r around a mass or charge has an area proportional to r^2; the force falls off as $1/r^2$ because at larger radii the lines of force are spread over a bigger area and their effect is diluted. If there were a fourth spatial dimension, the area of a sphere would be proportional to r^3 instead of r^2, and the force would follow an inverse cube law. The inverse square law is special because it allows orbits that are stable, in the sense that a planet's orbit only changes slightly in response to a slight "nudge" (such as would be caused by the recoil after an asteroid impact). Things would be catastrophically different if, instead, gravity obeyed an inverse cube law: a planet that was slightly slowed down would plunge into the Sun; and if it were slightly speeded up, it would spiral outward toward cold interstellar darkness. We would now interpret this conclusion as one of the special biophilic consequences of the three dimensions of space.

It was actually the eighteenth-century theologian William Paley—famous for his arguments that a universe that appears to be designed implies a Creator, just as a watch implies a watchmaker—who first noted the special stability of an inverse square law. (Paley had received a mathematical training in Cambridge in an era when Newtonian mechanics loomed large in the syllabus.) This realization buttressed his arguments for divine providence, but he did not relate the inverse square law to the number of dimensions of space. Had Paley been writing a century later, he would have applied a similar argument to atoms: electrons could not exist in stable "bound states" if electrical forces obeyed an inverse cube law.

Currently nothing seems absurd about a universe where space has extra dimensions: according to superstring theories, the ultra-early universe had ten or eleven. The extra ones would have wrapped up and "compactified" rather than expanded along with the others. If the extra dimensions were wrapped on the Planck scale, then they would not directly affect any experiment. But it is possible that the wrapping scale, though still microscopic, is not so tiny as the Planck length. The extra dimensions might then have some consequences that could be probed by particle physicists. Suppose, for instance, that two extra dimensions came into play on scales below 10^{-15} cm. Then Faraday's line-of-force argument tells us that the force within that radius would follow an inverse fourth power law rather than the usual inverse square. In three-dimensional space, gravity gets so fierce that quantum effects are important only on scales as small as the Planck length, 10^{-33} cm. But if gravity depended more steeply on the radius, going as an inverse fourth power rather than an inverse square, then quantum effects would set in before one reached down to a radius as tiny as 10^{-33} cm. The effective Planck length would no longer be quite so tiny, and the amount of compression needed to create a mini-black hole would be less extreme than in ordinary three-dimensional space. Some physicists speculate that small black holes could even be created in technically feasible accelerators.

Einstein famously spent his last thirty years seeking a unified theory of the physical laws. For him, it was a lone venture; but others, notably the English astrophysicist Arthur Eddington, were separately engaged in a similar quest. (Eddington, already celebrated for his classic and durable work on relativity and stellar structure, in later life became obsessed with a numerological "fundamental theory," according to which our universe was closed and finite. He even proposed a formula for the exact number of atoms in the universe.)*

These attempts were premature for many reasons. For instance, they focused on gravity and electromagnetism without taking cognizance of the other forces: the strong nuclear

---

*In his "Fundamental Theory," Eddington wrote: "I believe that there are "15,747,724,136,275,002,577,605,653,961,181,555,468,044,717,914,527, 116,709,366,231,425,076,185,631,031,296 protons in the Universe and the same number of electrons." (This number is actually 2^{256} x 136.) No living scientist believes this, and hardly any have made the effort to fathom Eddington's reasoning.

force and the so-called weak force, important in neutrinos and radioactivity. There have been other false dawns. But now, optimism is rife that superstring theory, or what is now called M-theory, offers the gateway to the fundamental equations. (Some specialists liken the present situation to the state of quantum theory before 1925, when everyone realized that a new paradigm was needed but there were only glimmerings of what it was.) Unified theories now engage young scientists, not just established dignitaries who can afford to risk over-reaching themselves and achieving nothing.

There is a daunting gap between the intricacies of ten or eleven dimensions and anything that we can observe or measure. It is unclear what determines the geometry of "ordinary" space: why did only three spatial dimensions expand to make our universe? Theorists can't yet tell us whether *all* the extra dimensions are wrapped up on microscopic scales, or whether there could be other universes separated from ours in a noncompactified extra dimension, just as many two-dimensional surfaces could exist, without contact with each other, in our three-dimensional space. (Whether a universe could have more than one *time* dimension is less straightforward. Certainly a language with more tenses would be needed to describe what happens in it.)

Generally, the mathematics needed by physicists can be taken "off the shelf." The non-Euclidean geometry that Einstein used to describe curved spaces had been developed by Riemann and others; the pioneers of quantum theory also found nineteenth-century mathematics sufficient for their purposes. But string theorists need twenty-first-century mathematics.

Superstring theory cannot yet account for various types of subatomic particles—quarks, gluons, and so forth—nor has it yet predicted something new, experimental or cosmological, caused by the extra dimensions. But many are already willing to bet on it partly because it almost seems to "predict" that a force like gravity should exist, but partly also for aesthetic reasons. There are precedents for this kind of stance. Einstein's theory of gravity—general relativity—gained widespread acceptance because of its aesthetic appeal, even when its empirical support was tenuous and imprecise. It transcended Newton's theory by offering deeper insights. Einstein accounted naturally, in a way that Newton did not, for why everything falls at the same speed and why gravity obeys an inverse square law. General relativity dates from 1916 and accounted for a long-standing anomaly in Mercury's orbit. It was famously corroborated by the measured deflection of starlight during an eclipse in 1919, but Einstein himself set more store by his theory's elegance: when asked how he would have reacted if the eclipse result had been discrepant, he replied that he would have been "sorry for the good Lord."

What a Fundamental Theory Won't Tell Us

A unified theory might, if it were achieved, genuinely be the greatest intellectual triumph of all time, certainly the culmination of an intellectual quest that started well before Newton and continued through Maxwell, Einstein, and their successors. It would elucidate the basic stuff that everything is made of and exemplify what the great physicist Eugene Wigner called "the unreasonable effectiveness of mathematics in the physical sciences." It would also illustrate the remarkable contingency—and it surely would be a contingency—that human mental powers could grasp the bedrock of physical reality.

But I hope it is not curmudgeonly to point out what it *would not* do. Such a theory would *not* signal the end of challenging science. Indeed, it would actually have minimal impact on most of science—even on most of cosmology. Two phrases often used in popular books—"final theory" or "theory of everything"—have connotations that are not only hubristic but very misleading.

Almost all scientists already believe that our everyday world is, in a sense, reducible to atomic physics. Just as every possible elliptical orbit of every planet is a solution of Newton's equations of motion, so every lump of material, whether living or inanimate, is governed by Schrödinger's equation— the basic equation of quantum theory which describes all atoms and assemblages of atoms. In practice, though, we can't solve this equation for anything more complicated than a single molecule. The complexity of the solution depends on how many atoms are involved, and also on the intricacy of their internal structure (for instance, a living cell is vastly more complex than a regular crystal made of the same number of atoms).

Moreover, even if we had a hypercomputer that *could* solve Schrödinger's equation for a complex macroscopic system and reproduce that system's behavior, the computer output would not yield any real insight. The insights that scientists seek require different concepts. For example, the flow of water—swirling eddies, breaking waves, and so forth—is best explained in terms of wetness, vorticity, and turbulence. These are emergent properties that exist only on scales large enough to allow us to treat a liquid as a continuum and ignore quantum efforts. Likewise, what goes on in a computer could be attributed to electrons moving in complicated circuits, but this notion misses the essence, the logic encoded in those signals. And, as Steven Pinker says, "Human behavior makes the most sense when it is explained in terms of beliefs and decisions, not in terms of volts and grams."

Each science, from chemistry to social psychology, has its own irreducible and emergent concepts and its distinctive types of explanation. There are general "laws of nature" in the macroscopic domain that are just as much of a challenge as anything in the microworld and are conceptually autonomous from it—for instance, there are general mathematical laws that describe the transition between regular and chaotic behavior and apply to phenomena as disparate as dripping taps and animal populations.[3]

The sciences are often likened to different levels of a building—logic in the basement, mathematics on the first floor, then particle physics, then the rest of physics and chemistry, and so forth as we climb upwards. But the analogy with a building breaks down because the superstructure (the "higher-level" science dealing with complex systems) is not imperiled by an insecure base.

Even if we knew the basic laws, we still would not understand how their consequences have unfolded over the last 13 billion years. We are baffled by complex patterns and interconnections—incomplete knowledge of the microworld generally is not the impediment.

This issue came into focus during the debates about the Superconducting Super-Collider (SSC)—a giant and hugely expensive accelerator, under construction in Texas but canceled by the U.S. Congress after $2 billion had already been spent on it. The hyperbole of its advocates led to a backlash from other scientists—even from some physicists who felt that complex materials, fluid flows, and chaos theory offered intellectual challenges as great as those in particle physics, and the resources consumed by the SSC would have been disproportionate.[4]

Nonetheless, as Steven Weinberg has emphasized, some sciences can claim a distinctive status, a special "depth."[5] If you go on asking why? why? why? you end up with a fundamental question either in particle physics or in cosmology: the sciences of the very small and the very large. This is an important feature of our universe. We seek unified theories of the cosmos and microworld not because the rest of science (or even the rest of physics) depends on them, but because they deal with deep aspects of reality.

The lure of the "final theory" is very strong. Ambitious students want to tackle the No. 1 challenge. But an undue focus of talent in one highly theoretical area is likely to be frustrating for all but a few exceptionally talented (or lucky) individuals. I advise my own students to multiply the importance of a problem by the small probability that they will solve it, and maximize that product. I remind them also of Peter Medawar's wise remark that "no scientist is admired for failing to solve a problem beyond his competence. The most he can hope for is the kindly contempt earned by utopian politicians."[6]

Despite Einstein's aphorism about the comprehensibility of the world that I quoted in the Prologue of this book, it would be astonishing if human brains were "matched" to all aspects of the external world. Some of nature's complexity may never be explained or understood.

11 Laws and Bylaws in the Multiverse

Many Universes?

I described in chapter 9 how the entire domain astronomers observe, extending at least 10 billion light-years, could have inflated from an infinitesimal speck; moreover, the inflationary growth could have led to a universe so large that its extent requires a million-digit number to express it. But even this vast expanse of space may not be everything there is: patches where inflation does not end may grow fast enough to provide the seeds for other Big Bangs. If so, our Big Bang wasn't the only one but may even be part of an eternally reproducing cosmos.

There are other conjectures that suggest a multiplicity of universes. For instance, whenever a black hole forms, processes deep inside it might trigger the creation of another universe into a space disjoint from our own. If that new universe were like ours, stars, galaxies, and black holes would form in it, and those black holes would in turn spawn another generation of universes, and so on, perhaps ad infinitum. Alternatively, if there were extra spatial dimensions that were not tightly rolled up, we may be living in one of many separate universes embedded in a higher-dimensional space.

All these theories are tentative and should be prefaced by something akin to a health warning. But they give us tantalizing glimpses of a dramatically enlarged cosmic perspective. The entire history of our universe could be just an episode, one facet, of the infinite multiverse. Were this indeed so, some features of our universe would be less surprising. Let me sketch why I think this is so.

A Special Recipe?

The distinctive details of our universe, and of everything in it (ourselves included), seem to be the outcome of what might be called an accident. The size and shape of our home Galaxy are the outcome of quantum fluctuations imprinted when the universe was the size of a golf ball; so is the layout of galaxies in the Local Group around us. The gases that ended up in our Sun had been, for billions of years, churned up by the shearing motions in our rotating Galaxy and buffeted by supernova explosions. Our Earth (along with the other inner planets, Mercury, Venus, and Mars) is an agglomeration of rocks and asteroids; the largest crash scooped out the material that made the Moon. Earth's surface has been molded by continental drift, by volcanism, and by further impacts. These and other terrestrial contingencies have controlled the topography and climate and determined the emergence and extinction of species. On a more parochial scale of space and time, each of us is the outcome of time and chance—the key events in the lives of all our ancestors. On a still smaller, microscopic, scale we owe our genetic inheritance to the near-random fate of individual spermatozoa.

Obviously, we can never explain all the contingencies that led from a Big Bang to our own birth here 13 billion years later. The outcome depended crucially on a recipe encoded in the Big Bang, and this recipe seems to have been rather special. I argued in chapter 5 that the emergence of such an intricate variety from a simple beginning does not conflict with any fundamental principle. But a degree of fine-tuning—in the expansion speed, the material content of the universe, and the strengths of the basic forces—seems to have been a prerequisite for the emergence of the hospitable cosmic habitat in which we live.

The following are some prerequisites for a universe containing organic life of the kind we find on Earth.

First of all, it must be very large in its spatial extent compared to individual particles, and very long lived compared with basic atomic processes. Indeed, this is surely a requirement not only for our universe, but for any hypothetical universe that a science fiction writer could plausibly find interesting. If atoms are the basic building blocks, then clearly nothing as elaborate as an ecosystem could be constructed unless there were huge numbers of them. Nothing much could happen in a universe that was was too short lived: an expanse of time, as well as space, is needed for evolutionary processes.

We have seen that a force such as gravity is crucial. But an interesting universe requires it to be very weak. If gravity were not exceedingly weak on the scale of atoms, then stars (gravitationally bound fusion reactors) would be small and short lived; gravity would crush anything larger than an insect, and there would be no time for complex evolution. Any interesting recipe must involve at least one very large number. This in itself is not fine-tuning—it is merely a constraint. And there is another constraint: the cosmic repulsion in empty space must be very weak (equivalently, the number lambda must be very small); otherwise, this disruptive force would have prevented gravitationally bound structures from forming.

But even if such structures form in a universe as large and long lived as ours, the outcome could be very boring: it could contain just black holes or inert dark matter, and no atoms at all. An interesting universe requires the kind of asymmetry in the laws that allows an excess of matter or antimatter, so that enough atoms can exist. Atoms need not be the *dominant* constituent in terms of mass: in our own universe the dark matter outweighs them by a factor of 5 to 10. But if there were, say, ten times fewer atoms than there actually are, they would remain in diffuse gas that would never condense into galaxies and stars.

The requirement of an interesting universe pins down other numbers in a specific narrow range. As discussed in chapter 5, the number Q, measuring the cosmic texture, cannot be too far from 1/100,000. If it were still smaller, the expansion could be so smooth—with no initial ripples or fluctuations—that no structures would develop. If Q were much larger, the universe would be so rough that it would collapse into huge black holes: an inclement environment for any form of life that we can readily imagine.

There is tuning in the microworld as well. The nuclear fusion that powers stars depends on the balance between two forces: the electrical repulsion between any two protons, and the strong countervailing nuclear force that attracts them to each other (and which also attracts the electrically neutral neutrons). The laws must not only allow protons and neutrons to exist, but they must allow the variety of atoms required for complex chemistry. If nuclear forces were slightly weaker, no chemical elements other than hydrogen would be stable: there would be no periodic table, chemistry would be a trivially simple subject, and there would be no nuclear energy to power the stars. But if the nuclear forces were slightly stronger than they actually are relative to electric forces, two protons could stick together so readily that ordinary hydrogen would not exist, and stars would evolve quite differently. Some of the details are still more sensitive. For instance, we noted in chapter 3 that carbon would not be so readily produced in stars were it not for some apparent fine-tuning in the properties of its nucleus, which depend even more sensitively on this same number.

What Does the Fine-Tuning Mean?

If our existence depends on a seemingly special cosmic recipe, how should we react to the apparent fine-tuning? We seem to have three choices: we can dismiss it as happenstance; we can acclaim it as the workings of providence; or (my preference) we can conjecture that our universe is a specially favored domain in a still vaster multiverse. Let's consider them in turn.

Happenstance (or Coincidence)

Perhaps a fundamental set of equations, which may some day be written on T-shirts, fixes all key properties of our universe uniquely. It would then be just an unassailable fact that these equations permitted the immensely complex evolution that led to our emergence.

But I think there would still be something to wonder about. It is not guaranteed that simple equations permit complex consequences. To take an analogy from mathematics, consider the beautiful pattern known as the Mandelbrot set. This pattern is encoded by a short algorithm but has an infinitely deep structure: tiny parts of it reveal novel intricacies no matter much they are magnified. In contrast, you can readily write down other algorithms, superficially similar, that yield very dull patterns. Why should the fundamental equations encode something with such potential complexity as our actual universe rather than the boring or sterile universe that many recipes would lead to?

One hard-headed response is that we could not exist if the laws had boring consequences. We manifestly are here, so there's nothing to be surprised about. But I am afraid this leaves me unsatisfied. I am impressed by a metaphor given by the Canadian philosopher John Leslie. Suppose you are facing a firing squad. Fifty marksmen take aim, but they all miss hitting you, the target. If they had not all missed, you would not have survived to ponder the matter. But you would not leave it at that: you would still be baffled, and you would seek some further reason for your luck. Likewise, we should surely probe deeper, and ask why a unique recipe for the physical world should permit consequences as interesting as those we see around us (and which, as a by-product, allowed us to exist).

Providence or Design

Design in the cosmos is the traditional theme of what used to be called "natural theology." Two centuries ago, William Paley introduced the famous metaphor of the watch and the watchmaker—adducing the eye, the opposable thumb and so on as evidence of a benign Creator. This line of thought fell from favor, even among most theologians, in post-Darwinian times. We now view any biological contrivance as the outcome of prolonged evolutionary selection and symbiosis with its surroundings.

As mentioned in chapter 10, Paley included the fact that gravity obeys an inverse square law among his design arguments. He actually could not muster much astronomical ammunition for his main theological thesis. He said, in his quaint way, that "astronomy is not the best medium through which to prove . . . an intelligent creator, but that, this being proved, it shows beyond all other sciences the magnificence of his operations."

Paley might have reacted differently, however, if he had known about the providential-seeming physics that led to galaxies, stars, planets, and the ninety-two elements of the periodic table (encapsulated in the fine-tuned strength of the nuclear force) and the number Q that imprints cosmic structure. And he would have been impressed by other seemingly biophilic features of basic physics and chemistry—for instance, those that give water its unusual properties of expanding when it cools and freezes.

These features can not be as readily dismissed as the old claims for design in living things. This is because the basic laws governing stars and atoms are a given, and nothing biological can react back on them to modify them. A modern counterpart of Paley is John Polkinghorne, a Cambridge physics professor who turned theologian in later life (and who was one of my own physics teachers). He interprets our fine-tuned habitat as "the creation of a Creator who wills that it should be so."[1]

A Special Universe Drawn from an Ensemble, or Multiverse

If one does not believe in providential design, but still thinks the fine-tuning needs some explanation, there is another perspective—a highly speculative one, so I should reitererate my health warning at this stage. It is the one I much prefer, however, even though in our present state of knowledge any such preference can be no more than a hunch.

There may be many "universes" of which ours is just one. In the others, some laws and physical constants would be different. But our universe would not be just a random one. It would belong to the unusual subset that offered a habitat conducive to the emergence of complexity and consciousness.

The analogy of the watchmaker could be off the mark. Instead, the cosmos may have something in common with an off-the-rack clothes shop: if the shop has a large stock, we are not surprised to find one suit that fits. Likewise, if our universe is selected from a multiverse, its seemingly designed or fine-tuned features would not be surprising.

This hypothesis may not seem "economical": indeed, it might seem to flout, absolutely maximally, the dictum now known as Ockham's Razor—the injunction of the fourteenth-century sage William of Ockham "not to multiply hypotheses more than necessary." At first sight, nothing seems more conceptually extravagant than invoking multiple universes. But this concept follows from several different theories (albeit all speculative), and opens up a new vision of our universe as just one atom selected from an infinite multiverse.*

Are Questions about Other Universes Part of Science?

Science is an experimental or observational enterprise, and it is natural to be troubled by assertions that cannot be checked empirically. Some might regard the other universes as being in the province of metaphysics rather than physics. But I

*There is a risk of semantic confusion here. The usual definition of "universe" is of course "everything there is." It would be neater to redefine the whole enlarged ensemble as "the universe" and then introduce some new term—for instance, "the metagalaxy"—for the domain that cosmologists and astronomers can directly observe. But so long as these concepts all remain so conjectural, it is best to leave the term "universe" undisturbed, with its traditional connotations, even though this then demands a new word, the "multiverse," for a (still hypothetical) ensemble of "universes."

think they already lie within the proper purview of science. It is not absurd or meaningless to ask, "Do unobservable universes exist?" even though no quick answer is likely to be forthcoming. The question plainly cannot be settled by *direct* observation, but relevant empirical evidence *can* be sought, which could lead to an answer.

There is actually a blurred transition between the readily observable and the absolutely unobservable, with a very broad gray area in between. To illustrate this, one can envisage a succession of four horizons (see fig. 11.1), each taking us farther than the last from our direct experience:

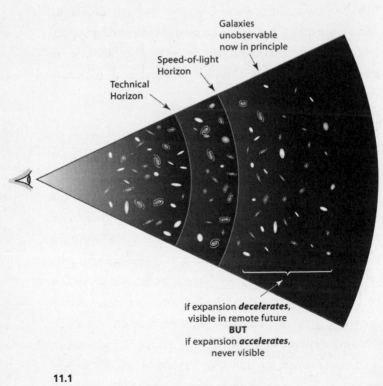

Galaxies
unobservable
now in principle

Speed-of-light
Horizon

Technical
Horizon

if expansion *decelerates*,
visible in remote future
BUT
if expansion *accelerates*,
never visible

11.1
Successive "credibility horizons" (see text for an explanation).

1. Limit of present-day telescopes

There is a limit to how far out into space our present-day instruments can probe. Obviously there is nothing fundamental about this limit: it is constrained by current technology. Many more galaxies will undoubtedly be revealed in the coming decades by bigger telescopes now being planned. We would obviously not demote such galaxies from the realm of proper scientific discourse simply because they have not been seen yet. When ancient navigators speculated about what existed beyond the boundaries of the then-known world, or when we speculate now about what lies below the oceans of Jupiter's moons, Europa and Ganymede, we are speculating about something "real"—we are asking a scientific question. Likewise, conjectures about remote parts of our universe are genuinely scientific, even though we must await better instruments to check them.

2. Limit in principle at present era

Even if there were absolutely no technical limits to the power of telescopes, our observations are still bounded by a horizon, set by the distance that any signal, moving at the speed of light, could have traveled since the Big Bang. This horizon demarcates the spherical shell around us at which the redshift would be infinite. There is nothing special about the galaxies on this shell, any more than there is anything special about the circle that defines your horizon when you are in the middle of an ocean. On the ocean, you can see farther by climbing up your ship's mast. But our cosmic horizon cannot be extended unless the universe changes, so as to allow light to reach us from galaxies that are now beyond it.

When our universe is, say, twice as old as it is now, this horizon will be twice as far away. But if that expansion is decelerating, then each galaxy, having slowed down, will be *less than* twice as far away, so the horizon of our remote descendants will also encompass extra galaxies that are beyond our horizon today. It is, to be sure, a practical impediment if we have to await a cosmic change taking billions of years, rather than just a few decades—maybe—of technical advance, before a prediction about a particular distant galaxy can be put to the test. But does that introduce a difference of principle? Surely the longer waiting time is a merely quantitative difference, not one that changes the epistemological status of these faraway galaxies.

3. Never-observable galaxies from "our" Big Bang

But what about galaxies that we can *never* see, however long we wait? In chapter 5 I discussed evidence that we inhabit an accelerating universe. As in a decelerating universe, there would be galaxies so far away that no signals from them could yet have reached us; but if the cosmic expansion is accelerating, we are now receding from these remote galaxies at an ever increasing rate, so if their light has not reached us yet, it never will. Such galaxies are not merely *unobservable in principle now*—they will be beyond our horizon *forever*. But if a galaxy is *now* unobservable, it hardly seems to matter whether it remains unobservable forever, or whether, as in a decelerating universe, it would come into view if we waited a trillion years. (And I have argued, under (2) above, that the latter category should certainly count as "real.")

4. Galaxies in disjoint universes

The never-observable galaxies in (3) would have emerged from the same Big Bang as we did. But suppose that, instead of causally disjoint regions emerging from a single Big Bang (via an episode of inflation), we imagine separate Big Bangs. Are space-times that are completely disjoint from ours any less real than regions that never come within our horizon in what we would traditionally call our own universe? Surely not. So these other universes should count as real parts of our cosmos, too.

This step-by-step argument (those who don't like it might dub it a slippery slope argument) suggests that whether other universes exist or not is a scientific question. So how do we answer it?

Scenarios for a Multiverse

Many scenarios could lead to multiple universes. Andrei Linde, Alex Vilenkin, and others have performed computer simulations depicting an "eternal" inflationary phase where many universes sprout from separate Big Bangs into disjoint regions of space-time. Alan Guth and Lee Smolin have, from different viewpoints, suggested that a new universe could sprout inside a black hole, expanding into a new domain of space and time inaccessible to us. And Lisa Randall and Raman Sundrum suggest that other universes could exist, separated from us in an extra spatial dimension. These disjoint universes may interact gravitationally, or they may have no effect whatsoever on one another. In the hackneyed analogy where the surface of a balloon represents a two-dimensional universe embedded in our three-dimensional space, these other universes would be represented by the surfaces of other balloons: any bugs confined to one, and with no conception of a third dimension, would be unaware of their counterparts crawling around on another balloon. Other universes would be separate domains of space and time. We could not even meaningfully say whether they existed before, after, or alongside our own because such concepts make sense only insofar as we can impose a single measure of time, ticking away in all the universes.

Guth and Edward Harrison have even conjectured that universes could be made in the laboratory by imploding a lump of material to make a small black hole. Is our entire universe perhaps the outcome of some experiment in another universe? Smolin speculates that the daughter universe may be governed by laws that bear the imprint of those prevailing in its parent universe. If so, the theological arguments from design could be resuscitated in a novel guise, further blurring the boundary between natural and supernatural phenomena.

Parallel universes are also invoked as a solution to some of the paradoxes of quantum mechanics, in the "many worlds" theory first advocated by Hugh Everitt and John Wheeler in the 1950s. This concept was prefigured by Olaf Stapledon, as one of the more sophisticated creations of his *Star Maker:* "Whenever a creature was faced with several possible courses of action, it took them all, thereby creating many ... distinct histories of the cosmos. Since in every evolutionary sequence of the cosmos there were many creatures and each was constantly faced with many possible courses, and the combinations of all their courses were innumerable, an infinity of distinct universes exfoliated from every moment of every temporal sequence."

None of these scenarios has been simply dreamed up out of the air: each has a serious, albeit speculative, theoretical motivation. However, one of them, at most, can be correct. Quite possibly none is: there are alternative theories that would lead just to one universe.

Firming up any of these ideas will require a theory that consistently describes the extreme physics of ultra-high densities, how structures on extra dimensions are configured, and so forth. But consistency is not enough: there must be grounds for confidence that such a theory is not a mere mathematical construct but applies to external reality. We would develop such confidence if the theory accounted for things we *can* observe that are otherwise unexplained. At the moment, we have an excellent framework called the "standard model" that accounts for almost all subatomic phenomena that have been observed. But the formulas of the standard model involve numbers, about eighteen altogether, which cannot be derived from the theory but have to be inserted from experiment. Any theory that gave some insight into why there are particular families of particles, and into the nature of the nuclear and electric forces, would acquire credibility; we would then be disposed to pay serious regard to other predictions it made, even if we could not directly test them.

Einstein's theory of gravity, or general relativity—dates from 1916. It took more more than fifty years before any tests could measure the distinctive effects of the theory with better than 10 percent accuracy. But now the scope and precision of empirical tests have improved so much, and yielded such com-

prehensive and precise support for Einstein, that it would require very compelling evidence indeed to shake our belief that
general relativity is the correct classical theory of gravity. In
consequence, we now have confidence in what the theory tells
us even about regions we cannot probe, such as the interiors of
black holes. Likewise, we take seriously our ideas about nuclear
reactions inside stars and in the hot early universe because
they are based on theories of atoms and their nuclei that have
been well confirmed experimentally.

Perhaps, in the twenty-first century, physicists will for-
mulate a theory that copes with an extrapolation right back to
the Planck time and earns our confidence by accounting for
hitherto unexplained phenomena accessible to experiment. If
such a theory were to predict many Big Bangs, then we would
have as much reason to believe in separate universes as we
now have for believing statements about black holes, or about
helium formation in the first few minutes after the Big Bang.
Some day we may therefore have grounds either for belief or
for disbelief in other universes.

Universal Laws, or Mere Bylaws?

If other universes exist, theory may also offer clues to a further
key question about them: How much variety do they display?
Some theorists, Frank Wilczek for instance, regard the question "Are the laws of physics unique?"—a less poetic para-
phrase of Einstein's question quoted in the Prologue—as a key
scientific challenge for the new century. If there were something uniquely self-consistent about the actual recipe, then any
Big Bang would trigger a universe that was just a rerun of ours.

But a far more interesting possibility (which is certainly tenable in our present state of ignorance of the underlying laws) is that *the underlying laws governing the entire multiverse may allow variety among the universes.* What we call the laws of nature govern the entire domain we observe, but they may in this grander perspective be *local bylaws*, consistent with some overarching theory governing the ensemble but not uniquely fixed by that theory. Many things in our cosmic environment—for instance, the exact layout of the planets and asteroids in our solar system—are accidents of history. Likewise, the recipe for an entire universe may be arbitrary.

The same balance between chance and necessity arises in biology. Our basic development—from embryo to adult—is encoded in our genes, but many aspects of our development are molded by our environment and experiences. There are far simpler examples of the same dichotomy—snowflakes, for instance. Their ubiquitous sixfold symmetry is a direct consequence of the properties and shape of water molecules. But their immense variety depends on their environment—on the fortuitous temperature and humidity changes during each flake's growth. If we had a fundamental theory, we would know which aspects of nature were direct consequences of the bedrock theory (just as the symmetrical template of snowflakes is due to the basic structure of a water molecule) and which are the outcome of accidents (like the distinctive pattern of a particular snowflake). The accidental features could be imprinted during the cooling that follows the Big Bang, rather as a piece of red-hot iron becomes magnetized when it cools down, but with an alignment that may depend on chance factors. They could have other contingent causes, such as the influence of another nearby universe separated from ours in a fifth dimension.

The cosmological numbers in our universe, omega, Q, and lambda, and perhaps some of the so-called constants of laboratory physics as well, could be arbitrary rather than uniquely fixed by some final theory. If so, then the off-the-rack clothes shop analogy—where there is a large stock of universes, as it were—would remove any reason for being surprised by the apparent fine-tuning of these numbers in our particular home universe.

Some features of our universe could then only be explained by "anthropic" argument (see fig. 11.2). Although this

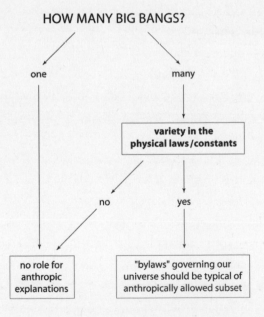

HOW MANY BIG BANGS?

one   many

variety in the
physical laws/constants

no   yes

no role for
anthropic
explanations

"bylaws" governing our
universe should be typical of
anthropically allowed subset

11.2
Flow chart illustrating how the status of anthropic explanations depends on the character of fundamental physical laws.

style of explanation raises hackles among some physicists, it is analogous to what observers or experimenters do when they allow for selection effects in their measurements: if there are many universes, most of which are not habitable, we should not be surprised to find ourselves in one of the habitable ones.

Testing Multiverse Theories Here and Now

We may one day have a convincing theory that tells us whether a multiverse exists and whether some of the so-called laws of nature are just parochial bylaws in our cosmic patch. But while we are waiting for that theory—and it could be a long wait—the "off-the-rack clothes shop" analogy can already be checked. It could even be refuted: this would happen if our universe turned out to be *even more specially* tuned than our presence requires.

To illustrate this line of reasoning, let's consider one seemingly tuned cosmic number: the energy latent in empty space which causes a cosmic repulsion and is measured by the number lambda. Physicists would expect lambda to be large because it is a consequence of a very complicated microstructure of space. A small lambda is, however, a prerequisite for our existence: an unduly fierce cosmic repulsion would disrupt galaxies. Perhaps there is only a rare subset of universes where lambda is below the threshold that allows galaxies and stars to form (as below the blurred line on fig. 11.3). Lambda in *our* universe obviously had to be below that threshold. But if our universe were drawn from an ensemble in which lambda was equally likely to take any value, we would not expect it to be *too far below it*.

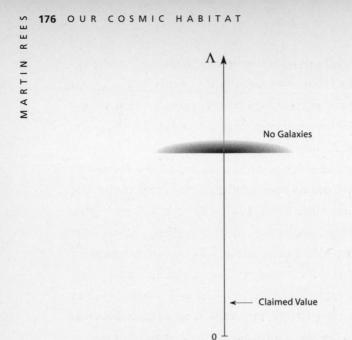

11.3
Constraints on lambda: if cosmical repulsion were
too strong, no galaxies would form. Is our universe
typical of anthropically allowed universes, or is
lambda far lower than our existence requires?

If we are indeed in an accelerating universe, as the current evidence suggests, the actual value is five to ten times
below that threshold. That would put our universe between
the tenth or twentieth percentile of universes in which galaxies could form. In other words, our universe is not signifi-
cantly more special, with respect to lambda, than our emergence demanded. But suppose that, contrary to current
indications, future observations showed that lambda made no
discernible contribution to the expansion rate and was *thousands of times* below the threshold, not just five to ten times.

This "overkill precision" would raise doubts about the hypothesis that lambda was equally likely to have any value and suggest that it was zero for some fundamental reason (or that it had a discrete set of possible values, and all the others were well above the threshold).[2]

I have taken lambda just as an example. We could analyze other important numbers of physics in the same way to test whether our universe is typical of the habitable subset that could harbor complex life. The methodology requires us to decide what values are compatible with our emergence. It also requires a specific theory that gives the probability of any particular value. For instance, in the case of lambda, are all values equally probable, or is there some more complicated formula? With this information, one can then ask if our actual universe is "typical" of the subset in which we could have emerged. If it is an atypical member even of this subset (not merely of the entire multiverse), then our hypothesis would be disproved.

As another example of how "multiverse" theories can be tested, consider Smolin's conjecture that new universes are spawned within black holes, and that the physical laws in the daughter universe retain a memory of the laws in the parent universe: in other words, there is a kind of heredity. Smolin's concept is not yet bolstered by any detailed theory of how any physical information (or even an arrow of time) could be transmitted from one universe to another. It has, however, the virtue of making a prediction about our universe that can be checked.

If Smolin were right, universes that produce many black holes would have a reproductive advantage, which would be passed on to the next generation. Our universe, if it is an outcome of this process, should therefore be near-optimum in its propensity to make black holes, in the sense that any slight tweaking of the laws and constants would render black hole formation less likely.[3] In our universe, black holes form as the endpoint of massive stars, and also in the centres of galaxies. It would then require only astronomical observations and an astrophysical understanding of these formation processes to test whether any change in the physics of atoms, nuclei, or galaxies would enhance the propensity to form black holes. I personally think Smolin's prediction is unlikely be borne out, but he deserves our thanks for presenting an example that illustrates how a specific multiverse theory can be vulnerable to disproof.

These examples show that some claims about other universes may be refutable, as any good hypothesis in science should be. We cannot confidently assert that there were many Big Bangs—we just don't know enough about the ultra-early phases of our own universe. But the physics of ultradense materials may, when applied to the Big Bang, predict multiple universes. Moreover, this same theory may tell us that each universe cools down differently, ending up with different expansion rates, contents, dimensionality, and microphysics.

Elucidating whether the underlying laws *are* as permissive as this is a challenge to twenty-first-century physicists. If things work out that way, then so-called anthropic explanations would become legitimate—indeed, they'd be *the only type of explanation we'll ever have* for some important features of our universe.[4] Efforts to seek fundamental formulas for some of the key numbers of physics would then be as misguided as Kepler's attempts to relate the sizes of planetary orbits to the Platonic solids (cubes, tetrahedra, and so forth).

A Seventeenth-Century Flashback

Kepler knew only about our solar system. Moreover, he thought that the orbits of the planets should be circles in exact mathematical ratios. Today we don't expect that. Our Earth traces just one ellipse out of an infinity of possibilities allowed by Newton's laws—the exact shape is a result of its complicated history and origins. Its orbit is special only insofar as it allows an environment conducive for evolution (not getting so close to the Sun that water boils, nor being so far away that it's perpetually frozen).

Perhaps our traditional perspective on the universe and the physical laws that govern it will go the way of Kepler's concept of Earth's orbit. What we have traditionally called "the universe" may be the outcome of one Big Bang among many, just as our solar system is merely one of many planetary systems in the Galaxy. Just as the pattern of ice crystals on a freezing pond is an accident of history rather than a fundamental property of water, so some of the seeming constants of nature may be arbitrary details rather than being uniquely defined by the underlying theory.

Our own universe—our cosmic habitat—has a simple recipe, but it isn't quite as simple as it might have been. It contains dark matter as well as atoms. As an extra complication, dark energy in empty space exerts a repulsion that overwhelms gravity on the cosmic scale. Some theorists are upset by these developments because they frustrate their craving for maximal simplicity. I think we can learn a lesson from the cosmological debates in the seventeenth century. Galileo and Kepler were upset that planets moved in elliptical orbits, not in perfect circles. But later Newton showed that all elliptical orbits could be understood by a single unified theory of gravity. Likewise, our universe may be just one of an ensemble of all possible universes, constrained only by the requirement that it allows our emergence. But to regard this outcome as ugly may be as myopic as Kepler's infatuation with circles. Newton was perhaps the greatest scientific intellect of the second millennium. Perhaps his third-millennium counterpart will uncover a mathematical system that governs the entire multiverse.

Finally, let us recall Hubble's words in his classic 1936 book, *The Realm of the Nebulae:* "Only when empirical resources are exhausted do we reach the dreamy realm of speculation." We still dream and speculate. But there has been astonishing empirical progress since Hubble's time owing to large telescopes on the ground, to the great instrument in space that bears his name, and to other technical advances.

There are three great frontiers in science: the very *big,* the very *small,* and the very *complex.* Cosmology involves them all. First, cosmologists must pin down the basic numbers such as Ω, and find what the dark matter is. I think there is a good chance of achieving this goal within ten years. Second, theorists must elucidate the exotic physics of the very earliest stages, which entails a new synthesis between cosmos and microworld. It would be presumptuous for me to place bets here. But cosmology is also the grandest environmental science, and its third aim is to understand how a Big Bang described by a simple recipe evolved, over 13 billion years, into our complex cosmic habitat: the filamentary layout of galaxies through space, the galaxies themselves, the stars, planets, and the prerequisites for life's emergence. No mystery in cosmology presents a more daunting challenge than the task of fully elucidating how atoms assembled—here on Earth and perhaps on other worlds—into living beings intricate enough to ponder their origins.

Scales of Structure

Our universe involves an immense range of scales—from smaller than atoms to larger than galaxies. We are each made up of between 10^{28} and 10^{29} atoms. The human scale is, in a numerical sense, poised midway between atoms and stars. It is actually no coincidence that nature attains its maximum complexity on this intermediate scale: anything larger, if it were on a habitable planet, would be vulnerable to breakage or crushing by gravity.

But even the nearest stars are millions of times farther away than the Sun. Our Galaxy contains a hundred billion stars altogether. There are at least as many galaxies within range of our telescopes as there are stars in a galaxy; altogether these visible galaxies contain 10^{78} atoms.

There are over 40 powers of 10 between the smallest known subnuclear particles probed by accelerators and the largest scales surveyed by telescopes. And the actual range of scales is surely far wider. Theorists discuss microscales 20 powers of 10 smaller than atomic nuclei—right down at the Planck length.

Figure A.1 depicts this range of scales on a diagram that

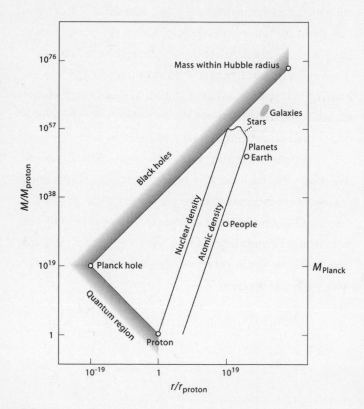

A.1

Scales of cosmic structure.

plots mass against radius on a logarithmic scale. The sizes of black holes are shown: they give a line of slope 1 on this logarithmic plot. Note that a minihole the size of a proton has a mass of about 10^{39} protons. Because gravity is so weak, that vast number of protons would have to be packed into the volume normally occupied by one proton before gravity overwhelms nuclear and electrical forces.

Ordinary solids—grains of sand, humans, and asteroids—lie on a line of slope 3, mass being proportional to the cube of radius. But when they get so massive that they contain about 10^{54} atoms (the mass of Jupiter) their own gravity starts to crush them. Above this mass range lie the stars.

A black hole weighing 10^{19} times more than a proton has a radius 10^{19} times less than proton. This radius equals the Planck length. For this hole, the uncertainty in its location due to quantum effects is as large as its entire size.

The largest scales (top right in the figure) correspond to our observable universe. It is, however, a theme of this book that there are intimate links between the very large and the very small—between inner space and outer space—and that our cosmos extends beyond even the largest scales depicted here by many more powers of ten than the entire range spanned by the diagram.

NOTES

Chapter 1   Planets and Stars

1. This particular quotation from Lord Kelvin (then William Thomson) is from *Macmillan's Magazine* (5 March 1862), p. 288; he repeated his argument in several later articles.

2. These instruments would consist of four or five telescopes in space, arrayed as an interferometer so that the light from the star itself cancels out by interference (the peaks of waves from one telescope neutralizing the troughs from the other) and won't drown out the light from any (far fainter) planet.

Chapter 2   Life and Intelligence

1. Conway-Morris's comments were made at a 2000 conference at the Pontifical Academy of Sciences. See also his book *The Crucible of*

Creation: The Burgess Shale and the Rise of Animals (Cambridge University Press, 1998).

Just as there can be debate about arbitrariness versus convergence in natural selection, so there is controversy about whether the chemical basis of other biospheres would be the same as that of life on Earth. Proteins are built up from about twenty different amino acids, and the number of possible proteins that are one hundred amino acids long—even if only a tiny fraction of them are water soluble and have surfaces that are chemically active—is surely vast compared to the number that actually exist in our biosphere. If that is indeed so, alien life—even if it is carbon-based like ours—could involve an immense variety of different biochemistries.

But some have argued that there is nonetheless just a limited range of possible chemical templates, just as there might be a limited range of body designs. The Harvard biologist George Wald (in *Origins of Life* 5 [1974]: 7) inferred this limitation from the properties of chlorophyll, the molecule that harnesses the light from the Sun in all green plants. This molecule is actually rather inefficient, in that it intercepts only a small fraction of the energy of sunlight, being insensitive to the part of the spectrum where the Sun's light is most intense. The very fact that this molecule has been chosen by natural selection on Earth, despite seeming to be suboptimal, led Wald to suspect that no better molecule exists. Chlorophyll could therefore, he suggests, be a generic feature of life around stars of any color.

2. Earthlike planets induce speeds of only about a centimeter per second. Such a slow crawl would be undetectable in an ordinary star. However, the movement is measured differently when the star is a pulsar. A clock is set by the pulse period. If the star gets nearer to us, the pulses arrive earlier: if it moves away, they arrive later. The arrival time can be measured to a precision far better than a millisecond—accurate enough to reveal the motions induced by orbiting planets smaller than Mercury.

3. The pulsar in the Crab Nebula emits visible light as well as radio waves and can be seen through any large telescope. But the pulse repetition rate, thirty pulses per second, is so high that the eye responds to it as a steady source. Had it been equally bright but spinning more slowly—say, ten times a second—the remarkable properties of the little

star in the Crab Nebula could have been discovered seventy years ago. How would the course of twentieth-century physics have been changed if superdense matter had been detected in the 1920s, before neutrons were discovered on Earth? One cannot guess, except that astronomy's importance for fundamental physics would surely have been recognized far sooner.

4. The amount of headstart is hard to assess. Some stars in our Galaxy formed 5 billion years before the Sun. But they would have been made from almost pristine material. The first planetary systems probably did not form until galactic oxygen and silicon had built up to at least a third of their abundance in our solar system. Nonetheless, there could have been planets around stars whose formation preceded the Sun by at least 2 billion years: longer than the time since the first multicellular organisms formed on Earth.

Chapter 5   Pregalactic History

1. The quote is from Lemaître's popular book, *The Primordial Atom* (Kluwer, 1953).

2. The easiest way to envision what happens to the radiation is to imagine the universe to be divided by a lattice of transparent walls into cubical cells. As the universe expands, the walls recede from one another and all the cubes expand. If the universe is homogeneous, then there is as much radiation entering any box (across each transparent wall) as is passing out of it. So the content of a box would be the same if the transparent walls were replaced by mirrors. We can then envision a simpler phenomenon: the expansion of a box—small on the cosmic scale—containing radiation that bounces off its reflecting walls. The properties of the universe determine the expansion rate, but otherwise the physics can be thought of in local terms.

3. One important clue relates to the proportions of helium and deuterium (heavy hydrogen) that emerge from the Big Bang. From observations we know the actual proportions of these atoms in the universe. The mix that would emerge from the Big Bang can be calculated and depends on the density of the atoms. Theory and observation are in gratifying agreement if there are as many atoms as we actually see in stars and gas. If, on the other hand, there were five or ten times more atoms in the universe than we actually see, the predicted mix emerging

from the Big Bang would be discrepant with observation. At higher densities the reactions would go faster: too much helium would be produced, and not enough deuterium (an intermediate product) would survive. Extra "exotic" particles that do not participate in nuclear reactions, however, can add mass without scuppering the concordance between the calculated and observed abundances of these types of atoms.

4. The Big Bang theory could have been observationally refuted in four other ways:

   a. According to the theory, the early universe would have contained neutrinos—elusive particles, with no electric charge, that are made in radioactive decays. Indeed, neutrinos and photons would have come into balance in the hot early phases: neutrinos would therefore, like photons, outnumber the ordinary atoms in the universe by a factor of about a billion. This means that they would be the dominant mass in the universe even if they weighted only a few billionths as much as a proton. Physicists have discovered that neutrinos may have a mass, but a very tiny one—probably so little that it would take 100 billion to make up the mass of a proton. They therefore would not be expected to make a significant contribution to the overall mass of the universe. But suppose the mass had turned out to be a millionth that of a proton—still tiny, but more than it actually seems to be. The Big Bang theory would then predict that neutrinos would contribute far too much dark matter. We would then have been forced to develop a different theory in which the background radiation was somehow generated much later in cosmic history, without concurrent neutrino production.

   b. As mentioned in the previous note, the observed amount of deuterium fits well with the amount predicted to survive from the Big Bang, provided that the density of atoms equals that in visible matter, and the dark matter is in some other form. However, if astronomers had discovered *far more* deuterium—for instance, if it was 1/5000 rather than 1/50,000 as abundant as ordinary hydrogen—then there would be a problem. Such a large amount could survive from the Big Bang only if the reactions converting it into helium proceeded more slowly. This would require *an even lower density of atoms than we actually observe* in stars and gas.

c. If the searches for angular fluctuations in the microwave background had failed to find anything and the radiation were say, ten times smoother than it was actually found to be, this would have implied a value of Q in the early universe that was too small to be compatible with the existence of present-day structures as large as clusters of galaxies.

d. According to the Big Bang theory, the temperature would have been higher in the past. When the light from the most distant observed galaxies set out, they were bathed in radiation that was at 15 degrees Kelvin (that is, 15 above absolute zero). Atoms and molecules in remote intergalactic clouds are observed to be heated to this temperature: if high-redshift gas had been found to be only at 3 degrees Kelvin, this would have refuted the theory.

Chapter 6   Black Holes and Time Machines

1. The key conceptual advances in gravitational physics since the 1960s can be traced to collaborations and interactions among a small number of leading workers. These people nearly all emerged from three research schools—those led by Yakov Zeldovich in Moscow, John Wheeler in Princeton, and Dennis Sciama (who was my own mentor, and that of Stephen Hawking and of many others who contributed to relativity theory) in Cambridge. The interactions among the three groups were almost universally cooperative and constructive. (*Black Holes and Time Warps* [W. H. Norton, 1994], a book by the American theorist Kip Thorne, gives an individualistic but insightful perspective on this research community.) In these respects it is atypical: science normally progresses in more boisterous and less coherent ways.

2. The quote is from J. A. Wheeler, *American Scientist* 56, 1 (1968).

3. Quoted from a 1975 lecture by S. Chandrasekhar, reprinted in *Truth and Beauty* (University of Chicago Press, 1987), p. 54.

4. Penrose's views on black holes and on other issues quoted here, are summarized in his book *The Emperor's New Mind* (Oxford University Press, 1988).

Chapter 7   Deceleration and Acceleration

1. The gravitational attraction depends not on density alone but on (density) + 3(pressure/c^2). If the universe were filled with material that had a large negative pressure (i.e., like a tension), then the second term can outweigh the first and cause a major qualitative change: rather than slowing down, the expansion actually accelerates. The following simple argument explains why the pressure associated with vacuum energy must be negative. If an ordinary gas expands by pushing a piston, then it ends up cooler than before—in other words, with less thermal energy. The difference represents the energy used up by pushing the piston. But if there were energy in a vacuum, then expanding the volume would create more energy, meaning that instead of the piston being pushed and gaining energy, it would have to *provide* energy. The piston therefore has to be pulled out, as though it were working against a tension or negative pressure. This counterintuitive result is important for an early inflationary phase of the universe as well as today if the dynamics of the universe is dominated by the energy of empty space (lambda) or by quintessence.

2. Physicists don't have a specific concept of quintessence but conceive it as an extra field pervading the universe. A more exotic idea is that quintessence is a manifestation of extra dimensions: if there were a parallel universe in another three-dimensional space separated from ours in an extra dimension, the gravitational interaction between them would generate a field with the properties of quintessence, its strength depending on the proximity of the two spaces.

3. A similar shift occurred with respect to the mixture of elements (as discussed in chapter 2). It was not known before the 1920s that only 2 percent of the atoms were in heavy elements. It took so long to appreciate this key fact because neither hydrogen nor helium reveals itself conspicuously in the light of stars, whereas stellar spectra show strong lines due to the other types of atoms. Also, of course, these volatile elements are underrepresented here on Earth and in the other inner planets.

Chapter 8   The Long-Range Future

1. The limit on information storage actually depends on the surface area. This dependence means that a black hole of a certain size con-

tains an amount of information that depends on the square of its mass. Similarly, the amount of information storable in an accelerating universe depends on the area of its horizon.

2. The quote is from an article by J. Ostriker and P. Steinhart in *Scientific American,* January 2001.

3. The paper is by W. Busza, R. L. Jaffe, J. Sandweiss, and F. Wilczek, *Reviews of Modern Physics* 72 (2000): 1125–37. See also A. Dar, A. de Rujula, and U. Heinz, *Phys. Lett.* B 470 (1999): 142–48. A thoughtful critique of the issues raised is given by F. Calogero in *Interdisciplinary Science Reviews* 25 (2000): 191–202. Greg Benford's novel *COSM* is on this theme.

Chapter 9   How Things Began

1. These heavy-ion experiments are the ones that have triggered the unease (mentioned in chapter 8) regarding the risk of a runaway conversion of everything into strange matter.

2. In the simplest situation, this would predict that the number Ω that measures the density of the universe should be exactly 1. As mentioned earlier, however, there appear to be only enough atoms and dark matter to make Ω around 0.3 . That's why the claim for a non-zero lambda, or for quintessence, was seized on so enthusiastically: these exotic forms of energy can make up the deficit and fulfill the flatness prediction.

Chapter 10   Cosmos and Microworld

1. Dirac's motive for conjecturing that G might be steadily decreasing in any case seems misguided from our modern perspective. As described in chapter 3, Robert Dicke realized that the lifetime, as well as the size, of stars was determined by the very large number that reflects the weakness of gravity on the atomic scale. He noted that we were not observing the universe at a random time. It was natural that we were on the scene during an era when stars had formed but had not all died—in other words, when the age of the universe was, very roughly, the same as a stellar lifetime. At that epoch, the Hubble radius would be the age of a typical star multiplied by the speed of light, and Dirac's approximate coincidence would be automatically satisfied. There is one crucially important large number in nature, 10^{39}, related to the weakness of gravity, and the others are straightforwardly linked to it.

2. The quantitative implications of the Oklo reactor for the constancy of various physical "constants" were reviewed and improved by F. J. Dyson and T. Damour in *Nuclear Physics* B480 (1996): 37–54. This article gives reference to earlier work.

3. A classic exposition of this viewpoint is Philip Anderson's "*More Is different,*" *Science* 177 (1972): 393–96.

4. The outcome of the ssc's cancelation is that CERN, the international laboratory in Geneva that has for decades been the focus for elementary particle research in Europe, is becoming a world laboratory, with major involvement by scientists in the United States, Japan, Russia, and elsewhere. This newly achieved European hegemony in particle physics contrasts with the traditional situation in space research, where Europe's effort is smaller-scale than NASA's, and the European Space Agency has generally been a minor partner in the biggest projects, such as the Hubble Space Telescope. Thus there is now—albeit by chance rather than proper planning—a benign transatlantic complementarity in the two "big sciences", space research and particle physics, where facilities are so expensive that duplication would be wasteful.

5. See Steven Weinberg's *Dreams of a Final Theory* (Pantheon, 1993).

6. In *Advice to a Young Scientist* by P. Medawar (Oxford University Press, 1979).

Chapter 11   Laws and Bylaws in the Multiverse

1. The quote is from John Polkinghorne, *Quarks, Chaos and Christianity* (SPCK Triangle Press, 1994).

2. The line of argument used here—that we expect our universe to be as special as our existence requires but not much more so—could have been used a century ago to refute Ludwig Boltzmann's "fluctuation hypothesis." Boltzmann worried about the arrow of time: he thought everything was a fluctuation in an infinite cosmos where overall there was no distinction between past and future. If he were right, then it would be far more likely to have a brief fluctuation that constituted one brain, complete with memories and sensations, rather than the whole cosmic tapestry itself. Unless we are solipsists, believing the vast external universe is an illusion, then the actual fluctuation seems vastly larger

(and more improbable) than our existence would necessarily entail. Such arguments would be fatal to Boltzmann's hypothesis.

3. Before Smolin's conjecture can be checked, he needs to make it more specific, particularly in spelling out the precise definition of "optimum." For instance, should the rate of formation, the space density, or the total number of holes in a universe be optimized?

4. By far the most comprehensive treatise on this subject is *The Anthropic Cosmological Principle* by J. D. Barrow and F. Tipler (Oxford University Press, 1986). The term "anthropic principle" is in my view an unfortunate one, with pretentious connotations. A better phrase is "anthropic reasoning." Many theorists disparage anthropic arguments as a copout or stop-gap—a way of allaying our curiosity when we have no "proper" explanation. Obviously we should strive to explain as many aspects of the physical world as we can in terms of fundamental formulas. But success is not guaranteed: indeed, some important numbers may be cosmic environmental accidents that have *no* fundamental explanation any more than the detailed shape of a particular snowflake does, or the exact orbits of planets.

INDEX

religious responses to, 125; on rotation, 142

Next Generation Space Telescope, 57

1984 (Orwell), 31

Novikov, Igor, 88, 95

nuclear fusion, 6

observability, 165–69, *166*

Ockham's Razor, 165

Oklo mine (Gabon, West Africa), 145–46

Omega (measurement of density), 104, 180, 193n.2

On the Infinite Universe and Worlds (Bruno), 19–20

Orion Nebula, 9

Orwell, George: *1984,* 31

OWL Telescope (OverWhelmingly Large Telescope), 57

oxygen, 16

ozone, 16

Paley, William, 149, 163–64

parallel universes. *See* multiverse

particle physics, 154, 194n.4

Payne, Cecilia, 37

Peebles, James, 67

Penrose, Roger, 90, 134

Penzias, Arno, 67, 68

periodic table of elements, 5–6, 44

51 Persei, 10

photons, 127–28, 190n.4

physical laws: as changing, 143–47, 144fn, 193n.1; as unique, 172–73

physics, 86, 154

Pinker, Steven, 154

Planck, Max, 83

Planck scales, 127, 147, 149–50, *184*

planets, 3–13; detection of, 9–12, 187n.2, 188n.2; Earthlike, 12–13, 20–21, 24, 188n.2; Jupiterlike, 13; orbits of, 124–25, 148–49, 178–79, 180; origins of, 8–10. *See also* solar system

Polkinghorne, John, 164

prediction, 99–104, *103*

pregalactic history, 65–86; dark matter in, 70–75, 189–90n.3; elements present in first few minutes, 69, *70*; eras of, 85; and gravity/complexity, 75–76, *77*, 78–80; gravity in, 71, 72; heat during the first few minutes, 69; radiation in, 67–68, *68*, 82–83, 189n.2, 191n.4; and the texture of the universe, 80–82, 82fn, 160–61. *See also* beginning of the universe; Big Bang

primeval atom. *See* Big Bang

proplyds (protoplanetary disks), 9–10, 125

protons, 45–46, 128–29

providence/design, 163–64

pulsars, 25, 188–89n.3

quantum theory, 5–6, 45–46, 127, 147, 151, 153

quantum vibrations/ripples, 134

quarks, 44, 120

Queloz, Didier, 10

quintessence, 108, 111, 119, 192n.2, 193n.2

radiation, 58–59; from black holes, 90–91; gravitational, 114–15; in pregalactic history, 67–68, *68*, 82–83, 189n.2, 191n.4

radioactivity, 150–51